Ergebnisse der exakten Naturwissenschaften

Herausgegeben von G. Höhler

Unter Mitwirkung von S. Flügge,
F. Hund und F. Trendelenburg

Schriftleitung E. A. Niekisch

37. Band

**Springer-Verlag
Berlin Heidelberg GmbH**

Anschrift des Herausgebers

Prof. G. Höhler, Institut für Theoretische Kernphysik der Technischen Hochschule, 75 Karlsruhe, Hauptstraße 54

Anschrift der Schriftleitung

Dr. E. A. Niekisch, Kernforschungsanlage Jülich, Arbeitsgruppe Institut für Technische Physik, 517 Jülich, Postfach 365

ISBN 978-3-662-15912-5   ISBN 978-3-540-37138-0 (eBook)
DOI 10.1007/978-3-540-37138-0

© by Springer-Verlag Berlin Heidelberg 1965
Originally published by Springer-Verlag Berlin Heidelberg New York in 1965
Softcover reprint of the hardcover 1st edition 1965

Library of Congress Catalog Card Number 25-9130

Brühlsche Universitätsdruckerei Gießen

Inhaltsverzeichnis

Paramagnetische Felder am Kernort

W. Donner und G. Süßmann

Institut für Theoretische Physik
der Universität Frankfurt a. M.

Eingegangen am 13. Juni 1964

Inhaltsverzeichnis

Summary

A theory of the paramagnetic correction of an external magnetic field at the site of the nucleus is developed. This effect is important especially for the experimental method of perturbed angular correlations, by which magnetic moments of excited nuclei are measured. It occurs if an "inner" electron shell is not closed, e. g. for the Lanthanons ($4f$-shell) or the Actinons ($5f$-shell).

The ion in view, considered to be isolated, is described by a single particle model with central field (e. g. Hartree's self consistent field), so that l is a good quantum number. Russell-Saunders coupling is assumed. For the ground multiplet L and S are evaluated by Hund's rule. The excited multiplets may be disregarded as they are not occupied at usual temperatures. Saturation, diamagnetic and other nonlinear effects in the magnetic field strength are neglected. Corrections to this model are discussed in the last section.

In the L-S-scheme, the strengths of the multiplet splitting and of the induced magnetic field are determined by two coefficients η and ξ, which depend on the electron configuration l^q. A detailed derivation of the expressions $\eta(l, q)$ and $\xi(l, q)$ is presented. At last, the quantum mechanical and thermodynamical average of the paramagnetic field in question is calculated.

1. Einleitung

Zur Messung von magnetischen Kernmomenten (Kopfermann 1956) benötigt man die genaue Kenntnis des magnetischen Feldes am Ort des Atomkerns (Abragam 1961). Dieses Feld ($\boldsymbol{B} + \boldsymbol{b}$) unterscheidet sich von dem von außen angelegten Magnetfeld ($\boldsymbol{B}$) um die diamagnetischen und paramagnetischen Effekte der Hüllenelektronen. (Vom Ferromagnetismus sehen wir ab, da wir uns in dieser Arbeit auf diejenigen Fälle beschränken wollen, in denen die jeweiligen Atome bzw. Ionen als frei betrachtet werden können.) Der Diamagnetismus ist so schwach, daß er bei der gegenwärtigen Meßgenauigkeit vernachlässigt werden kann. Der Paramagnetismus hingegen spielt immer dann eine wichtige Rolle, wenn die „inneren", d. h. chemisch oder optisch inaktiven Elektronen einen von Null verschiedenen Gesamtdrehimpuls besitzen, so daß schon für $\boldsymbol{B} = 0$ ein nicht verschwindendes magnetisches Moment resultiert. Das ist dann der Fall, wenn die inneren Elektronenschalen nicht voll abgeschlossen sind, also für die sog. Übergangselemente einschließlich der Lanthaniden und Aktiniden. In diesen Bereichen des Systems der chemischen Elemente werden nämlich die in Anbetracht ihrer hohen Energie verhältnismäßig nahe am Kern gelegenen d- und f-Schalen aufgefüllt.

Der Effekt ist für die f-Schalen bedeutend stärker und auch weniger von äußeren Einflüssen abhängig als für die d-Schalen; denn f-Elektronen sind im Mittel viel näher am Kern als d-Elektronen vergleichbarer Energie (Goeppert-Mayer 1941). Die quantitative Durchführung dieser Überlegung wird übrigens zeigen, daß es auf die Mittelwerte $\langle r^{-3}\rangle$ und $\langle [Z - z(r)] r^{-3}\rangle$ ankommt, wobei $z(r)$ die Anzahl der in einer Kugel vom Radius r enthaltenen, die Kernladungszahl Z abschirmenden Elektronen ist. Dies kann man sich folgendermaßen veranschaulichen: Das Magnetische Moment hat für Elektronen verschiedener Schalen die gleiche Größenordnung, und das induzierte Magnetfeld im Inneren der Schale ist größenordnungsmäßig gleich der mittleren Magnetisierung, dem Quotienten aus dem magnetischen Moment und dem Volumen der Schale.

Der in diesem Artikel behandelte Effekt ist besonders wichtig für die Messung der magnetischen Momente angeregter Atomkerne[*] mit

---

[*] Einen zusammenfassenden Bericht über diesen Gegenstand gab Bodenstedt (1962).

Hilfe von γ-γ-Winkelkorrelationen, die durch Larmorpräzession in einem äußeren Magnetfeld modifiziert werden*. Die untersuchten Atomkerne gehören Ionen an, die sich in Lösung befinden; man nimmt an, daß die Störfelder der umgebenden Ionen in ihrer Wirkung hinreichend schnell weggemittelt werden. Es liegen also ähnlich einfache Verhältnisse vor wie beim gewöhnlichen „makrophysikalischen Paramagnetismus", der vor allem von VAN VLECK (1932) untersucht worden ist. Die hier dargelegte, kompliziertere Theorie des „Paramagnetismus am Kernort" ist diesem Vorbild nachgebildet. Als erste haben GOLDRING und SCHAREN-BERG (1958) eine vereinfachte Version mitgeteilt, die in einem Teil der Anwendungsfälle eine brauchbare Näherung darstellt. Das genauere Resultat wurde von KANAMORI und SUGIMOTO (1958) angegeben und insbesondere auf Sm^{+++} angewandt.

Die Theorie beruht auf folgenden Näherungsannahmen, die recht gut begründet sind und ein wohlbestimmtes Modell ergeben:

a) Das betreffende Atom bzw. Ion wird als isoliert betrachtet, d. h. die Wirkungen benachbarter Atome bzw. Ionen werden vernachlässigt.

b) Für die Elektronenhülle des Ions setzen wir ein Einteilchen-Modell mit zentralsymmetrischem Potential voraus, so daß der Bahndrehimpuls l eine gute Quantenzahl ist. Konfigurationsmischungen werden also nicht berücksichtigt. Ferner nehmen wir an, daß $l \neq 0$ ist.

c) Dieses Zentralpotential denken wir uns in selbstkonsistenter** Weise bestimmt, allerdings wie üblich (HARTREE 1954) mit den folgenden vereinfachenden Annahmen:

α) Austauscheffekte werden vernachlässigt. Das heißt, bei der Bestimmung des selbstkonsistenten Abschirmungsfeldes wird auf die Diagonalisierung der $\binom{N}{2}$ Permutationsoperatoren $\mathscr{P}_{ij}$ verzichtet (N ist die Anzahl der Hüllenelektronen). Die Antimetrisierung der Zustandsfunktion, also der Übergang von einem Produkt orthogonaler Funktionen zu einer Slater-Determinante, wird erst nachträglich vorgenommen.

β) Drehimpulseffekte werden vernachlässigt. Das heißt, bei der Bestimmung des selbstkonsistenten Abschirmungsfeldes wird auf die Diagonalisierung des Gesamtdrehimpulses J verzichtet ($J = \Sigma j_i$ und $J^2 = J^2 + J$). Die Linearkombination der Slater-Determinanten zu Eigenfunktionen von J^2 wird erst nachträglich vorgenommen.

Von diesem Hartreeschen Modell*** unterscheidet sich das noch einfachere Schalenmodell der Atomhülle dadurch, daß die geringfügige Abhängigkeit des Abschirmungsfeldes von der abgeschirmten Elektronenbahn nicht in Rechnung gestellt wird.

---

* Über diese Meßmethode ist kürzlich ein Konferenzbericht erschienen (KARLSSON, MATTHIAS und SIEGBAHN 1964).

** Natürlich impliziert die bereits in b) geforderte Zentralsymmetrie des Abschirmungsfeldes ähnlich wie die Annahmen c α) und c β) gewisse Inkonsistenzen des Feldes, und zwar bezüglich der nicht abgeschlossenen und daher polarisierenden Elektronenschalen. Für $N \gg 1$ ist der hierdurch begangene Fehler jedoch klein.

*** Zur Hartree-Fockschen Näherung kommt man bekanntlich, wenn man die Austauscheffekte berücksichtigt, also allein mit der Vereinfachung β).

d) Das Magnetfeld $B = |\boldsymbol{B}|$ wird nur in erster störungstheoretischer Näherung berücksichtigt. Das ist erlaubt, weil B sehr viel kleiner als die mittlere elektrische Feldstärke am Ort der für den Paramagnetismus verantwortlichen Elektronen ist. Hierdurch entfällt insbesondere der Diamagnetismus.

e) Mit Ausnahme der Spin-Bahn-Kopplung werden die relativistischen Effekte vernachlässigt.

f) Es wird Russell-Saunders-Kopplung vorausgesetzt, d. h. L und S werden als gute Quantenzahlen angesehen.

g) Die L- und S-Werte des Grundmultipletts werden nach den Hundschen Regeln bestimmt. Die angeregten Multipletts werden als energetisch so hoch angenommen, daß die thermische Agitation sie bei normalen Temperaturen praktisch nicht erreichen kann.

h) Thermische Sättigungseffekte werden vernachlässigt, d. h. wir beschränken uns auf Terme, die in $\mu_e B / kT$ linear sind.

Von diesen Voraussetzungen sind b), d), g) und h) *wesentliche* Grundlagen der hier abgehandelten Theorie, während die übrigen durch einen bedeutend geringeren Arbeitsaufwand erweitert werden könnten. Die damit zusammenhängenden Korrekturen werden im letzten Abschnitt dieses Berichts diskutiert.

Die Theorie ist bisher nur auf f-Schalen angewandt worden. Als konkretes Beispiel wählen wir im folgenden stets die dreifach ionisierten Lanthaniden, bei denen die $4f$-Schale aufgefüllt wird. Die abgeleiteten Formeln gelten aber unter den angegebenen Voraussetzungen auch für andere Ionen oder Atome.

2. Die Schalenstruktur

Im folgenden Bericht wird überall die Schalenstruktur der Elektronenhülle vorausgesetzt. Wir beginnen daher mit einer kurzen Zusammenfassung der hierfür wesentlichen Begriffe. Dabei vernachlässigen wir einfachheitshalber die durch die Fockschen Gleichungen (Fock 1930) erfaßten Austauscheffekte, begnügen uns also mit den Hartreeschen Gleichungen (Hartree 1957). Wie üblich nehmen wir ferner kugelsymmetrische Ladungsverteilungen an.

Es bedeute q_{nl} die Anzahl, $r^{-1} P_{nl}(r)$ die Radialfunktion und ε_{nl} das Energieniveau der in der Unterschale (n, l) befindlichen Elektronen. Ferner sei $e^2 V_{nl}(r)$ die potentielle Energie eines solchen Elektrons im Feld des Atomkerns und aller übrigen Elektronen. Bei gegebenen $V_{nl}(r)$ sind dann $P_{nl}(r)$ und ε_{nl} durch die Differentialgleichung

$$\frac{\hbar^2}{2m_e} P_{nl}''(r) = \left[e^2 V_{nl}(r) + \frac{\hbar^2 l(l+1)}{2m_e r^2} - \varepsilon_{nl} \right] P_{nl}(r) \qquad (2.1)$$

mit den Randbedingungen

$$P_{nl}(0) = P_{nl}(\infty) = 0 \,, \qquad (2.2)$$

der Anordnung

$$\varepsilon_{l+1,\,l} < \varepsilon_{l+2,\,l} < \varepsilon_{l+3,\,l} < \cdots \tag{2.3}$$

und der Normierung

$$\int_0^\infty \mathrm{d}r\, P_{nl}^2(r) = 1 \tag{2.4}$$

bis auf ein unwichtiges Vorzeichen eindeutig definiert*. Nach dem Pauliprinzip muß stets $q_{nl} \leq 2(2l+1)$ sein. Das Komplement $\bar{q}_{nl} \equiv 4l + 2 - q_{nl}$ ist die „Anzahl der Löcher" in der Unterschale (n, l). Die Gesamtheit der q_{nl} wird als *Konfiguration* der Elektronenhülle bezeichnet; wir halten sie im folgenden stets fest.

Dann ist umgekehrt jedes der Potentiale $-eV_{nl}(r)$ bis auf eine belanglose additive Konstante durch die Ladungsdichten $-eP_{n'l'}^2(r)$ aller Atomelektronen bestimmt. Zunächst berechnet man auf Grund der Gleichung

$$4\pi r^2 \varrho_{nl}(r) = \sum_{n'=0}^{\infty} \sum_{l'=0}^{n'-1} (q_{n'l'} - \delta_{nn'}\, \delta_{ll'})\, P_{n'l'}^2(r) \tag{2.5}$$

die das Kernfeld abschirmende Ladungsverteilung** $\bar{e}\,\varrho_{nl}(r)$. Die innerhalb einer Kugel vom Radius r enthaltene effektive Ladung beträgt $eZ + \bar{e}\,z_{nl}(r)$, wobei Z die Ordnungszahl und

$$z_{nl}(r) = \int_0^r 4\pi r'^2\, \mathrm{d}\, r'\, \varrho_{nl}(r') \tag{2.6}$$

die „Abschirmungszahl" ist. Nach einem bekannten elektrostatischen Satz läßt sich hieraus leicht die effektive elektrische Feldstärke $+eV_{nl}'(r)$ berechnen

$$V_{nl}'(r) = \frac{Z - z_{nl}(r)}{r^2}. \tag{2.7}$$

Durch Integration dieser Gleichung erhält man schließlich

$$V_{nl}(r) = -\frac{Z - z_{nl}(r)}{r} + \int_r^\infty \mathrm{d}r'\, \frac{z_{nl}'(r')}{r'} = -\frac{Z}{r} + 4\pi \int_0^\infty \frac{\mathrm{d}r'\, \varrho_{nl}(r')\, r'^2}{\mathrm{Max}\,(r,\, r')}. \tag{2.8}$$

Die Zahl N aller Elektronen ist durch die Gleichungen

$$\sum_n \sum_l q_{nl} = N = 1 + z_{nl}(\infty) \tag{2.9}$$

gegeben. Der Ionisierungsgrad ist $Z - N$. Im Falle der dreiwertigen Lanthaniden ist dies natürlich gleich 3.

Wir können annehmen, daß dieses System von Differential- und Integralgleichungen bei gegebener Konfiguration genau eine Lösung besitzt. Es gibt danach nur einen konsistenten Satz von Funktionen $P_{nl}(r)$, $\varrho_{nl}(r)$, $z_{nl}(r)$, $V_{nl}(r)$. Wir beschränken uns im folgenden auf den Grundzustand und die niedrigsten Anregungen und somit auf die *Grundkonfiguration*. Diese ist dadurch definiert, daß die ungestörte

---

* Den Unterschied zwischen der Elektronenmasse m_e und der reduzierten Masse $m \approx m_e(1 - m_e/A\,m_p)$ vernachlässigen wir.

** Mit e wird die positive Elementarladung bezeichnet; die Elektronenladung ist also $-e$, wofür wir gelegentlich auch $\bar{e}$ schreiben werden.

Energie

$$E_0 = \sum_n \sum_l q_{nl} \left\{ \varepsilon_{nl} - \frac{e^2}{2} \int\limits_0^\infty \mathrm{d}\,r \cdot P_{nl}^2(r) \left[V_{nl}(r) + \frac{Z}{r} \right] \right\} \qquad (2.10)$$

ihren kleinstmöglichen Wert annimmt.

(Doch braucht diese Größe im folgenden nicht berechnet zu werden.) Hiernach gibt es höchstens eine Unterschale (n, l), die weder ganz leer noch ganz voll, d. h. für die $0 < q_{nl} < 4l + 2$ ist. Wir können daher für dieses eine q_{nl} einfach q schreiben und entsprechend auch $\bar q \equiv \bar q_{nl} = 4l + 2 - q$ setzen. Speziell haben wir es mit der Konfiguration $[Xe] (4f)^q$ zu tun, wobei $[Xe]$ gleich

$$(1s)^2 (2s)^2 (2p)^6 (3s)^2 (3p)^6 (4s)^2 (3d)^{10} (4p)^6 (5s)^2 (4d)^{10} (5p)^6 \qquad (2.11)$$

ist, d. h. $n = 4$, $l = 3$ und $q = q_{4f} = 14 - \bar q$ sowie $N = Z - 3 = 54 + q$. Die zur „ungestörten" Hüllenenergie (2.10) gehörigen Wellenfunktionen sind Linearkombinationen der Produkte

$$\Phi(r_1 \vartheta_1 \varphi_1 s_1, \ldots, r_N \vartheta_N \varphi_N s_N) = \prod_{j=1}^N r_j^{-1} P_{n_j l_j}(r_j)\, Y_{l_j m_j}(\vartheta_j\, \varphi_j)\, \delta_{\sigma_j s_j}, \qquad (2.12)$$

die zu der herausgegriffenen Konfiguration gehören, also für alle n und l den Bedingungen

$$q_{nl} = \sum_{j=1}^N \delta_{n_j n}\, \delta_{l_j l}$$

genügen. Man hat diejenigen Superpositionen der Produkte (2.11) aufzusuchen, die in allen N Elektronen antimetrisch sind und die zu scharfen Werten von J^2 und J_z gehören. Es kommen hier vor allem zwei Möglichkeiten in Betracht: Einmal Linearkombinationen von Slater-Determinanten und zum anderen Summen von Produkten zweier Determinanten, von denen die eine eine reine Ortsfunktion mit scharfem L^2 und L_z, die andere eine reine Spinfunktion mit scharfem S^2 und S_z ist. Im folgenden werden wir den zweiten Satz von ungestörten Wellenfunktionen benutzen[*].

3. Die Russell-Saunders-Kopplung

Unter den die Einteilchenbahnen verkoppelnden „Restwechselwirkungen" spielen die elektrostatischen die größte Rolle. Es handelt sich dabei um die zwischen den Elektronen wirkenden Coulomb-Kräfte, soweit sie nicht schon durch das mittlere Abschirmungsfeld berücksichtigt sind.

Da diese Kräfte spinunabhängig sind, lassen sie den Gesamtbahndrehimpuls $L = \Sigma l_j$ und die Spinsumme $S = \Sigma s_j$ der Elektronenhülle zeitlich konstant. Die verschiedenen Orientierungsmöglichkeiten eines Spinzustandes mit festem S relativ zu einem Bahnzustand mit festem L werden durch den Betrag J des Gesamtdrehimpulses $J = L + S$ charakterisiert. Die Gesamtheit der $2 \operatorname{Min}(L, S) + 1$ Zustände mit $J = |L - S|$, $\ldots$, $L + S$ bildet ein bestimmtes *Multiplett* mit den Quantenzahlen L

---

[*] Der erste Satz wird für die sog. *jj*-Kopplung benötigt. Diese tritt fast nur bei den schwersten Atomen (den Aktiniden) auf.

und S. Diese Kennzeichnung (^{2S+1}L) ist jedoch nur für $q \leq 2$ oder $\bar{q} \leq 2$ eindeutig: Hat man es mit mehr als zwei Teilchen und Löchern zu tun, so muß man zur Charakterisierung eines Multipletts der gegebenen Elektronenkonfiguration $(n, l)^q$ außer ^{2S+1}L im allgemeinen noch weitere Quantenzahlen mitteilen (beispielsweise Youngsche Diagramme, vgl. Anhang 1).

Da wir uns für den Grundzustand und die niedrigsten (nämlich thermisch) angeregten Niveaus der Elektronenhülle interessieren, müssen wir wissen, welches der vielen darstellungstheoretisch möglichen Multipletts am stärksten gebunden wird. Dieses Problem wird durch die zwei *Hundschen Regeln* (HUND 1927) in einfacher Weise und *eindeutig* gelöst. Diese besagen: 1. Die Spinzahl S nimmt den größtmöglichen der mit dem Pauliprinzip verträglichen Werte an. 2. Der Gesamtbahndrehimpuls L nimmt den größtmöglichen der mit dem Pauliprinzip verträglichen Werte an. In Formeln

$$\left.\begin{array}{l} S = \dfrac{1}{2}q\,;\, L = \dfrac{1}{2}q(2l+1-q) \text{ für } 0 \leq q \leq 2l+1 \\[2mm] S = \dfrac{1}{2}\bar{q}\,;\, L = \dfrac{1}{2}\bar{q}(2l+1-\bar{q}) \text{ für } 2l+1 \geq \bar{q} \geq 0 \end{array}\right\} q + \bar{q} \equiv 4l+2\,. \quad (3.1)$$

Die Gültigkeit der ersten Regel läßt sich leicht einsehen: Bei maximalem S ist die Spinfunktion $X^S(\sigma_1, \ldots, \sigma_q)$ der Unterschale maximal symmetrisch, ihre Bahnfunktion $\Phi^L(r_1, \ldots, r_q)$ daher nach dem Pauliprinzip maximal antisymmetrisch. Das aber hat zur Folge, daß die Elektronen einander nach Möglichkeit ausweichen, so daß die abstoßenden Restpotentiale in diesem Zustand schwächer zum Zuge kommen als in anderen. Für die zweite Regel ist uns eine einfache theoretische Begründung nicht bekannt. Wir berufen uns auf ihre empirische Gültigkeit, die von VAN VLECK (1932) auch für die Seltenen Erden festgestellt worden ist. In Tab. 1 findet man in den Spalten 3 und 4 die nach (3.1) berechneten Quantenzahlen S und L für alle 15 dreifach ionisierten Lanthaniden.

Tabelle 1

Z	Ion	S	L	J	ξ	g_J	$g_{(LS)J}$
57	La^{3+}	0	0	0	—	—	—
58	Ce^{3+}	1/2	3	5/2	0,04450	6/7	0,8560
59	Pr^{3+}	1	5	4	0,00741	4/5	0,8054
60	Nd^{3+}	3/2	6	9/2	0,00135	8/11	0,7307
61	Pm^{3+}	2	6	4	—0,00101	3/5	0,6027
62	Sm^{3+}	5	5/2	5/2	—0,00296	2/7	0,2928
63	Eu^{3+}	3	3	0	—0,00741	—	—
64	Gd^{3+}	7/2	0	7/2	0,04450	2	1,9938
65	Tb^{3+}	3	3	6	—0,00741	3/2	1,4918
66	Dy^{3+}	5/2	5	15/2	—0,00296	4/3	1,3231
67	Ho^{3+}	2	6	8	—0,00101	5/4	1,2370
68	Er^{3+}	3/2	6	15/2	0,00135	6/5	1,1926
69	Tm^{3+}	1	5	6	0,00741	7/6	1,1629
70	Yb^{3+}	1/2	3	7/2	0,04450	8/7	1,1412
71	Lu^{3+}	0	0	0	—	—	—

4. Die Spin-Bahn-Wechselwirkung

Die Russell-Saunders-Kopplung wird gestört (oder gelockert) durch Kräfte, die von den Spins der Elektronen abhängen. Es handelt sich hierbei um zwei verschiedene relativistische Effekte: (I) Ein mit der Geschwindigkeit v im elektrischen Feld E bewegtes Elektron* erfährt mit seinem magnetischen Moment μ das Magnetfeld $B' = -c^{-1} v \times E$. (II) Durch die Beschleunigung a erfährt das Elektron an seinem Spin s die Thomas-Präzession $\omega' = \frac{1}{2} c^{-2} v \times a$. Diese kinematischen Effekte sind mit folgender magnetisch-kinetischer Energie verknüpft:

$$H^{ls} = -\mu B' - \hbar s \omega .$$

Mit $m_e a = \bar{e} E = -e^2 V'(r) r/r$ und $r \times m_e v = \hbar l$ erhält man

$$H^{ls} = (g_e - 1) \frac{\hbar^2 e^2}{2 c^2 m_e^2} \frac{1}{r} V'(r) \, l \, s . \tag{4.1}$$

Hier bedeutet g_e den Landé-Faktor des freien Elektrons, d. h. $\mu/s = -g_e \mu_e$, wobei $\mu_e \equiv e \hbar/2 c \, m_e$ das (Bohrsche) Elektronen-Magneton ist**. Aus Messungen[a] und quantenelektrodynamischen Berechnungen[b] ist der Wert $g_e = 2{,}002_3$ bekannt. Um die Spin-Bahn-Energie H^{ls} der gesamten Atomhülle zu erhalten, muß man die in (4.1) angegebene Größe über alle N Elektronen summieren. Dabei wird aus Symmetriegründen $\Sigma \, l_i s_i = 0$ für alle vollbesetzten Unterschalen $\varepsilon_{n'l'}$ (d. h. für $q_{n'l'} = 4l' + 2$), so daß man nur das eine teilweise besetzte Niveau $\varepsilon_{nl} \equiv \varepsilon$ zu berücksichtigen braucht.

$$H^{LS} = \sum_{i=1}^{N} H_i^{ls} = \zeta \sum_{k=1}^{q} l_k \, s_k , \quad \text{mit} \quad \zeta = 2 \mu_e^2 (g_e - 1) \left\langle \frac{Z - z(r)}{r^3} \right\rangle . \tag{4.2}$$

Dabei ist $z \equiv z_{nl}$ nach (2.6) definiert. Die Mittelungsvorschrift $\langle \, \rangle$ bezieht sich auf die Wahrscheinlichkeitsverteilung der (n, l)-Unterschale (speziell der $4f$-Unterschale), lautet also folgendermaßen

$$\langle f(r) \rangle = \int_0^\infty dr \, P_{nl}^2(r) \, f(r) . \tag{4.3}$$

Dabei ist im Sinne des Schalenmodelles angenommen, daß die Quantenzahlen n und l trotz der elektrostatischen und der spinabhängigen Störungen hinreichend gut erhalten bleiben, um als wohldefinierte Konstanten behandelt werden zu können.

Da außerdem angenommen werden kann, daß die elektrostatische Energie H^C der zentralen Restwechselwirkungen groß ist gegen die Energie H^{LS} der spinabhängigen Kräfte, bleibt auch die Multiplettstruktur annähernd erhalten. Genauer: die Vektoren L und S werden

---

* $-\mu_e/\hbar \equiv -e/2c \, m_e$ ist Einsteins gyromagnetisches Verhältnis.

** Vom Standpunkt des Atomkerns aus gibt es kein Magnetfeld; dafür wird durch die Bewegung das elektrische Moment $c^{-1} v \times \mu$ induziert.

[a] A. Petermann: Helv. Phys. Acta 30, 407 (1957); C. M. Sommerfield: Phys. Rev. 107, 328 (1957); vgl. auch E. R. Cohen und J. W. M. Du Mond: Phys. Rev. Letters 1, 291 (1958).

[b] W. E. Lamb und R. C. Retherford: Phys. Rev. 72, 241 (1947); S. Triebwasser, E. S. Dayhoff und W. E. Lamb: Phys. Rev. 89, 98, 106 (1953).

variabel und ihre z-Komponenten Λ und Σ unbestimmt, aber ihre Beträge L und S behalten in erster Näherung ihre bestimmten und zeitlich konstanten Werte. Aus Symmetriegründen wird dann $\Sigma\, l_k\, s_k$ proportional zu $\boldsymbol{L\,S}$, wobei der Koeffizient η davon abhängt, welches Multiplett vorliegt. Nach Annahme der Hundschen Regeln ist das Multiplett des Grundzustandes wohldefiniert, und man kann η leicht berechnen (vgl. Anhang 2)

$$H^{LS} = \eta\,\zeta\,\boldsymbol{L\,S} \quad \text{mit} \quad \eta = \begin{cases} \dfrac{1}{2S} & \text{für} \quad 0 < q < 2l+1 \\[2mm] -\dfrac{1}{2S} & \text{für} \quad 2l+1 > \bar{q} > 0\,. \end{cases} \tag{4.4}$$

Für $q = 2l+1 = \bar{q}$ ist $H^{LS} = 0$ wegen $\boldsymbol{L} = 0$, so daß η bedeutungslos wird.

Diese Kopplung von $\boldsymbol{L}$ und $\boldsymbol{S}$ aneinander bewirkt eine Aufspaltung des betreffenden Multipletts in seine Komponenten-Niveaus, welche nur noch die $(2J+1)$-fache Orientierungsentartung besitzen. Auf Grund der Identität $2\boldsymbol{LS} = \boldsymbol{J}^2 - \boldsymbol{L}^2 - \boldsymbol{S}^2$ erhält man bekanntlich den Ausdruck

$$E^J = E^0 + \frac{1}{2}\,\eta\,\zeta\,J\,(J+1) \tag{4.5}$$

für die Energieaufspaltung. Der Grundzustand hat $J = |L - S|$ oder $J = L + S$ je nachdem, ob die Schale weniger bzw. mehr als halb voll ist (weil dann $\eta\,\zeta$ positiv bzw. negativ ist). Für die genau halbvolle Unterschale (Gd^{+++}) besteht das Grundmultiplett nur aus einem Niveau mit $J = S = l + \frac{1}{2}\,(= \frac{7}{2})$, vgl. (7.10).

Der zu E^J gehörige Zustand wird durch die übliche Drehimpulsaddition gebildet

$$\left|\begin{matrix} J \\ M \end{matrix}\right\rangle = \sum_{\Lambda = L}^{-L} \sum_{\Sigma = S}^{-S} \Phi_\Lambda^L X_\Sigma^S \begin{pmatrix} L & S & J \\ \Lambda & \Sigma & M \end{pmatrix}; \quad M = J, \ldots, -J. \tag{4.6}$$

Für $\Lambda + \Sigma \neq M$ muß natürlich der Clebsch-Gordan-Koeffizient $\begin{pmatrix} L & S & J \\ \Lambda & \Sigma & M \end{pmatrix}$ verschwinden, da wir die Quantisierungsachsen der drei Drehimpulse $\boldsymbol{L}$, $\boldsymbol{S}$ und $\boldsymbol{J}$ bequemlichkeitshalber zusammenfallen lassen. Die Funktionen $\Phi_\Lambda^L(\boldsymbol{r}_1, \ldots, \boldsymbol{r}_q)$ und $X_\Sigma^S(\sigma_1, \ldots, \sigma_q)$ sind zu den betreffenden Symmetriecharakteren kombinierte Produkte von Einteilchenzuständen der einen Unterschale (n, l). Die (durch Youngsche Diagramme charakterisierten) Austauschsymmetrien von Φ und X müssen zueinander adjungiert sein, damit Ψ voll antimetrisch wird*. Im Grundzustand ist, wie bereits gesagt wurde, Φ maximal symmetrisch und X maximal antisymmetrisch. Für $q = 2l+1$ ist Φ sogar voll symmetrisch und X voll antisymmetrisch, und für $\bar{q} \leq 2l+1$ gilt eine entsprechende Aussage für die Elektronenlöcher.

Die durch (4.4) zum Ausdruck gebrachte Kopplung bewirkt für festes L, S und J eine (reguläre) Präzession der Unterschale, bei welcher die Vektoren $\boldsymbol{L}$ und $\boldsymbol{S}$ mit der (konstanten) Winkelgeschwindigkeit

---

* Vergl. Anhang 1.

$\Omega^{LS} = \hbar^{-1}\eta\zeta J$ umlaufen, s. Abb. 1. Es gelten nämlich im Heisenbergbild die Bewegungsgleichungen $\dot{L} = -\dot{S} = \hbar\,\eta\,\zeta\,(S \times L)$. Hiernach sind die Skalare L^2, S^2 und der Vektor J konstant, ebenso die Parallelkomponenten $L_{||}$ und $S_{||}$, während die zu J orthogonale Komponente $L_\perp = -S_\perp$ mit der Kreisfrequenz

$$\Omega^{LS} = \hbar^{-1}\,\eta\zeta\sqrt{J(J+1)}$$

um J präzediert. Dies wird durch die Termabstände bestätigt: nach (4.5) ist $\Delta E^J \equiv \tfrac{1}{2}(E^{J+1} - E^J) = \eta\cdot\zeta(J+\tfrac{1}{2})$, also tatsächlich korrespondenzmäßig $= \hbar\Omega^{LS}$. Die stationären Zustände (4.6) werden durch die Präzession in sich übergeführt; insbesondere verschwinden die Erwartungswerte von $\dot{L}$ und $L_\perp$. Die Realität der Präzession zeigt sich aber beispielsweise darin, daß die Quadrate $(\dot{L})^2$ und $L_\perp^2$ oder auch das Spatprodukt $[\dot{L}, L, S]$ nichtverschwindende Erwartungswerte besitzen. Für die nichtstationären Linearkombinationen der $\left|{J\atop M}\right.$ kann man die Präzession bereits am Erwartungswert von L sehen.

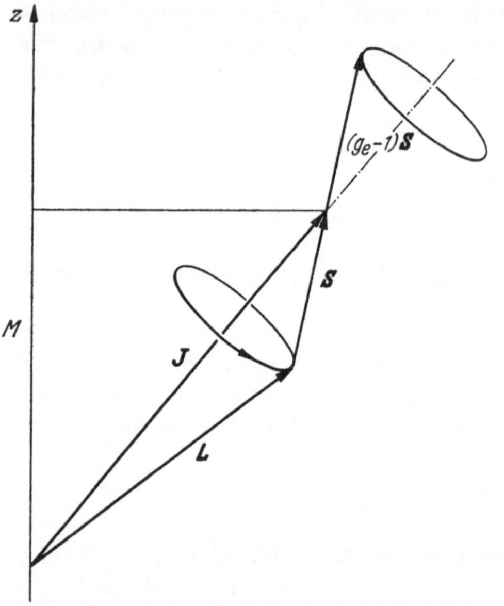

Abb. 1. Die beiden Präzessionen der Drehimpulsvektoren

Neben der bisher betrachteten Spin-Bahn-Kopplung gibt es auch eine magnetische Wechselwirkung der Elektronenspins. Sie ist mindestens um einen Faktor der Größenordnung $Z - z$ kleiner. Hinzu kommt, daß die Wechselwirkungsenergie H^{ss} für verschiedene Lagen der Elektronen bei festgehaltenen Spinrichtungen verschiedene Vorzeichen hat und sich daher im Mittel über die Richtungen weitgehend weghebt. Neue qualitative Effekte wie Termaufspaltungen werden nicht bewirkt*.

5. Die anomale Larmorpräzession (Zeeman-Effekt)

Als nächstes müssen wir der Tatsache Rechnung tragen, daß die Kugelsymmetrie des Problems durch das von außen angelegte homogene Magnetfeld B gestört wird: J wird variabel, lediglich die zu B parallele Komponente J_z bleibt konstant. Das Magnetfeld wirkt auf die Bahnen und auf die Spins der Elektronen (die Wirkung auf den Atomkern

---

* Die entsprechende magnetische Wechselwirkung der Bahn-Momente ist bereits in der kinetischen Energie $\Sigma\,p_i^2/2\,m_e$ enthalten.

können wir hier außer Betracht lassen). Die Bahnstörungen werden durch das zylindersymmetrische Vektorpotential $A = \frac{1}{2} B \times r$ ausgedrückt: Es gilt die verallgemeinerte de-Broglie-Relation $m_e v = \hbar\, k - (\bar{e}/c)\, A$ mit $k = -i\partial/\partial r$. Hiernach unterscheidet sich der kinetische Bahndrehimpuls $d = r \times m_e v$ vom kanonischen Bahndrehimpuls $\hbar l = \hbar (r \times k)$ um den „potentiellen Drehimpuls" $r \times (\bar{e}/c)\, A = \frac{1}{2} c^{-1} e\, r \times (r \times B)$. Während also l (in Übereinstimmung mit den Vertauschungsrelationen $l \times l = i l$) die bekannten ganzzahligen Eigenwerte besitzt, sind die Eigenwerte des „wirklichen" Bahndrehimpulses d *keine* ganzen Vielfachen von $\hbar$ mehr. Der Unterschied hängt mit der Larmorpräzession und dem Diamagnetismus der Atomhülle zusammen.

Der zuletzt genannte Effekt ist bereits in der ungestörten Wellenfunktion (4.6) enthalten. Dies erkennt man an der elektrischen Stromdichte

$$j(r) = \bar{e} \sum_{j=1}^{N} \frac{1}{2} \{v_j\, \delta(r - r_j) + \delta(r - r_j)\, v_j\} - $$
$$- g_e\, \mu_e \frac{\partial}{\partial r} \times \sum_j s_j\, \delta(r - r_j)\,,$$

die offenbar einen B-proportionalen Anteil $\bar{e} \sum (\omega_L \times r_j) \cdot \delta(r - r_j)$ enthält*. Für das magnetische Moment

$$m = \frac{1}{2}\, c^{-1} \int \mathrm{d}r^3\, r \times j$$

erhält man dementsprechend $m = (\bar{e}/2c\, m_e) \sum (d_i + g_e\, \hbar s_i)$ oder

$$m = -\mu_e(\hbar^{-1} D + g_e\, S) = -\mu_e(L + g_e\, S) - \beta\, B\,. \tag{5.1}$$

Dabei bedeutet $D = \sum d_i$ den „physikalischen" Gesamtdrehimpuls und

$$\beta = \frac{e^2}{4 c^2 m_e} \sum_{i=1}^{N} (r_i^2 - r_i \cdot r_i) \approx \frac{e^2}{6 c^2 m_e} \sum_{i=1}^{N} r_i^2 \tag{5.2}$$

die (diamagnetische) Magnetisierbarkeit der Atomhülle. Sie ist streng genommen ein symmetrischer Tensor, der jedoch in guter Näherung (nach Mittelung über viele Atome sogar genau) mit einem Skalar übereinstimmt. Solange die Elektronen als punktförmig angesehen werden können, trägt ihr Spin nichts zum Diamagnetismus bei. Im Gegensatz zum diamagnetischen Moment $- \beta\, B^2$ verschwindet das paramagnetische Moment $-\mu_e(L + g_e S)$ für die abgeschlossenen Schalen aus Symmetriegründen vollständig. (Beim diamagnetischen Moment annulliert die Kugelsymmetrie lediglich die auf B senkrechten Komponenten.)

Mit ihrem magnetischen Moment erfährt die Atomhülle das Drehmoment $m \times B$. Die hierzu und zur diamagnetischen Induktion gehörige Kopplungsenergie H^B ist kinetischer Natur, nämlich die Differenz zwischen der vollen kinetischen Energie $\sum (\frac{1}{2} m_e v_i^2 + g_e \mu_e\, s_i B)$ und der kanonisch-kinetischen Energie $\sum \hbar^2 k_i^2/2m$. Sie lautet

$$H^B = \mu_e(L + g_e\, S)\, B + \frac{1}{2} B\, \beta\, B = -m\, B - \frac{1}{2} B\, \beta\, B\,. \tag{5.3}$$

---

* $\omega_L \equiv -\bar{e} B/2c\, m_e$ ist die „normale" Larmorfrequenz.

Der diamagnetische Term $\frac{1}{2} \boldsymbol{B} \beta \boldsymbol{B}$ ist positiv definit, und zwar gleich $(e^2/4 c^2 m_e)\ \Sigma (\boldsymbol{B} \times \boldsymbol{r}_i)^2$. Aus Symmetriegründen darf er durch $\frac{1}{2} \beta B^2$ approximiert werden. Der Diamagnetismus ist ohnehin eine meist vernachlässigbare Korrektur, solange das Magnetfeld klein gegen *

$$3c\ \hbar/N e\ a_{\mathrm{H}}^2 \cong \frac{3}{N}\ \frac{c\ e^3 m_e^2}{\hbar^3} \cong 10^8\ \mathrm{Gs}$$

bleibt, was in praxi immer der Fall ist. Wir lassen den Diamagnetismus im folgenden außer Betracht. Das gilt auch für die entsprechende Korrektur am Wechselwirkungsterm (4.1).

Auch in bezug auf den Paramagnetismus werden wir das Magnetfeld als schwach behandeln und nur in der ersten störungstheoretischen Ordnung berücksichtigen. Die Krümmung der Magnetisierungskurve und andere Sättigungserscheinungen werden also ignoriert. Insbesondere soll die Frequenz ω_L der Larmor-Präzession klein sein gegen die Frequenz Ω^{LS} der Spin-Bahn-Präzession. Solange

$$B \ll \frac{J(Z-z)}{S}\ \frac{4\mu_e^2}{\hbar\ a_{\mathrm{H}}^3}\ \frac{c\ m_e}{e} = \frac{J(Z-z)}{S}\ \frac{e^7 m_e^2}{c\ \hbar^5} \cong (Z-z)\ 10^5\,\mathrm{Gs} \cong 10^6\,\mathrm{Gs}$$

bleibt, ist das erlaubt (z ist eine mittlere Abschirmungszahl). Die Störung (5.3) führt den Zustand (4.6) in den Zustand

$$\Psi_M^J = \sum_{J'=|L-S|}^{L+S} \left|{J' \atop M}\right) \left[\delta_{J'J} + \frac{1-\delta_{J'J}}{E^{J'}-E^J} \left({J' \atop M}\Big|m\Big|{J \atop M}\right) \boldsymbol{B} \right] \tag{5.4}$$

über. Dabei hat J_z seinen wohlbestimmten Wert M behalten, während $|\boldsymbol{J}|$ nicht mehr ganz scharf gleich $\sqrt{J(J+1)}$ ist. In (5.4) sind — außer den in $\boldsymbol{B}$ quadratischen Termen — die Einflüsse der höheren Multipletts vernachlässigt. Das ist im Rahmen der Russell-Saunders-Näherung ganz konsequent und wäre nur dann ungenau, wenn ein Multiplett mit den gleichen Symmetriecharakteren L und S zufällig sehr nahe an das Grundmultiplett herankäme. Der Erfolg der Van Vleckschen Theorie des gewöhnlichen Lanthaniden-Paramagnetismus (VAN VLECK, 1932) zeigt aber, daß solch ein unglücklicher Zufall offenbar nicht auftritt. Normalerweise hat die Voraussetzung $\eta\zeta \ll e^2/a_{\mathrm{H}}$ zur Folge, daß die Anregungsenergien der konkurrierenden Multipletts sehr viel größer als die in (5.4) vorkommenden Energienenner $E^{J'}-E^J$ sind.

Berechnet man den Erwartungswert $\langle A\rangle_M^J = \langle \Psi_M^J|\,A\,|\Psi_M^J\rangle$ irgend eines hermiteschen Operators A im Zustand (5.4), so erhält man

$$\langle A\rangle_M^J = \left({J \atop M}\Big|A\Big|{J \atop M}\right) + \boldsymbol{B} \sum_{J'}{}' 2\cdot \mathrm{Re}\ \frac{\left({J' \atop M}\Big|m\Big|{J \atop M}\right)\left({J \atop M}\Big|A\Big|{J' \atop M}\right)}{E^{J'}-E^J}\ . \tag{5.5}$$

Hierin deutet der Strich in Σ' an, daß der Summand $J' = J$ wegzulassen ist.

---

* $a_{\mathrm{H}} = \hbar^2/e^2 m_e$ = Bohrscher Radius. Bei der obigen Abschätzung ist angenommen, daß r^2 im Mittel die Größenordnung von a_{H}^2 hat.

Die M-Abhängigkeit des in (5.5) benötigten Diagonalelements von $\boldsymbol{m}$ ergibt sich aus dem Wigner-Eckart-Theorem zu

$$\begin{pmatrix} J' \\ M \end{pmatrix} m_\mu \begin{pmatrix} J \\ M \end{pmatrix} = \frac{(-1)^{J'+1-J}}{\sqrt{2J'+1}} \begin{pmatrix} J' \\ M \end{pmatrix} \begin{pmatrix} 1 & J \\ \mu & M \end{pmatrix} (J' \| \boldsymbol{m} \| J) \,, \tag{5.6}$$

wenn m_μ die μ-te sphärische Komponente von $\boldsymbol{m}$ ist ($\mu = 1, 0, -1$). Eine ähnliche Umformung gestattet natürlich auch das Diagonalelement von A, falls dieser Operator als Komponente eines Drehtensors* − beispielsweise eines Skalars oder eines Vektors − aufgefaßt werden kann. Damit (5.6) nicht verschwindet, müssen die Dreieckungleichungen $|J - 1| \leqq J' \leqq J + 1$ erfüllt sein, woraus u. a. folgt, daß J' und J sich höchstens um eins unterscheiden dürfen. Darüber hinaus muß natürlich die Gleichung $M = \mu + M$ erfüllt, also $\mu = 0$ sein.

Um das reduzierte Matrixelement $(J' \| \boldsymbol{m} \| J)$ auszuwerten, wird man (5.1) und den Ausdruck (4.6) mit einem möglichst bequemen M (etwa $= J$ oder $= 0$) in die linke Seite von (5.6) einsetzen. Das Resultat dieser elementaren Rechnung findet man z. B. bei VAN VLECK (1932). Eine ebenso elementare, nur in der Symbolik systematischere Ableitung bietet EDMONDS (1957, S. 111). Danach gilt bei Vernachlässigung des diamagnetischen Gliedes

$$\frac{(-1)^{L+S}(J' \| \boldsymbol{m} \| J)}{\mu_e \sqrt{(2J'+1)(2J+1)}}$$
$$= (-1)^J \begin{Bmatrix} J' & 1 & J \\ L & S & L \end{Bmatrix} (L \| \boldsymbol{L} \| L) + g_e (-1)^{-J'} \begin{Bmatrix} J' & 1 & J \\ S & L & S \end{Bmatrix} (S \| \boldsymbol{S} \| S) \,. \tag{5.7}$$

Die Racah-Koeffizienten $\begin{Bmatrix} J' & 1 & J \\ a & b & a \end{Bmatrix}$ sind von EDMONDS (1957, S. 130) angegeben**. Nach EDMONDS (1957, S. 76) ist ferner

$$(j \| \boldsymbol{j} \| j) = \sqrt{j(j+1)(2j+1)} \tag{5.8}$$

für jeden Drehimpuls j, und zwar in Verbindung mit seinen Eigenräumen $|j\rangle$; also insbesondere für $\boldsymbol{L}$ mit Φ^L, für $\boldsymbol{S}$ mit X^S und für $\boldsymbol{J}$ mit $|J\rangle$ oder Ψ^J. Durch Einsetzen erhält man unter Verwendung der Abkürzungen

$$\begin{cases} g_J = 1 + (g_e - 1) \cdot \dfrac{J(J+1) - L(L+1) + S(S+1)}{2J(J+1)} & \text{und} \\[2mm] h_J = \dfrac{(J+L+S+1)(J+L-S)(J-L+S)(-J+L+S+1)}{J} \end{cases} \tag{5.9}$$

---

* Drehtensor = irreducible tensor; vgl. z. B. FANO u. RACAH: "Irreducible Tensorial Sets" (1959) oder EDMONDS (1957), Kap. 5.1.

** Es gilt:
$$\begin{Bmatrix} J' & 1 & J \\ a & b & a \end{Bmatrix} = \begin{Bmatrix} J & 1 & J' \\ a & b & a \end{Bmatrix} = \begin{cases} \alpha_J(a, b) & \text{für} \quad J' = J \\ \beta_J(a, b) & \text{für} \quad J' = J - 1 \end{cases}$$

mit
$$\alpha_J(a, b) \equiv (-1)^{J+a+b+1} \cdot 2 \cdot \frac{J(J+1) + a(a+1) - b(b+1)}{\sqrt{2a(2a+1)(2a+2) \cdot 2J \cdot (2J+1)(2J+2)}} = \alpha_a(J, b)$$

und
$$\beta_J(a, b) \equiv (-1)^{J+a+b} \cdot \sqrt{\frac{2(J+a+b+1)(J+a-b)(J-a+b)(-J+a+b+1)}{2a(2a+1)(2a+2)(2J-1)2J(2J+1)}} \,.$$

Insbesondere ist $\beta_{|a-b|}(a, b) = \beta_{a+b+1}(a, b) = 0$.

die reduzierten magnetischen Momente

$$\begin{cases} (J\| \boldsymbol{m}\|J) = -\mu_e\sqrt{J(J+1)(2J+1)}\,g_J\,, \\ (J-1\|\boldsymbol{m}\|J) = -(J\|\boldsymbol{m}\|J-1) = \mu_e(1-g_e)\frac{1}{2}\sqrt{h_J}\,. \end{cases} \tag{5.10}$$

Für $|J'-J| \geqq 2$ und $J' = J = 0$ ist $(J'\|\boldsymbol{m}\|J) = (J\|\boldsymbol{m}\|J') = 0$.
Wegen

$$\begin{pmatrix} J & 1 & J \\ M & \mu & M \end{pmatrix} = -\delta_{\mu 0}\frac{M}{\sqrt{J(J+1)}} \tag{5.11}$$

wird

$$\begin{pmatrix} J \\ M \end{pmatrix} m_\mu \begin{pmatrix} J \\ M \end{pmatrix} = -\mu_e M g_J \delta_{\mu 0}\,. \tag{5.12}$$

Danach kann $g_J \cdot J$ als die Komponente von $\boldsymbol{m}$ in Richtung $\boldsymbol{J}$ in dem — durch die Quantenzahlen (L, S) genauer charakterisierten — Niveau $|J)$ aufgefaßt werden, so daß g_J nichts anderes ist als der Landésche Faktor dieses Niveaus.

Dies wird bestätigt, wenn man den Energiewert $E_M^J \equiv \langle H\rangle_M^J$ berechnet, indem man in (5.5) für A die Gesamtenergie $H(\equiv E)$ einsetzt. Zunächst gilt nach (5.3) in der angenommenen Näherung

$$\begin{pmatrix} J \\ M \end{pmatrix} H \begin{pmatrix} J' \\ M \end{pmatrix} = E^J \cdot \delta_{JJ'} - \boldsymbol{B}\begin{pmatrix} J \\ M \end{pmatrix} \boldsymbol{m} \begin{pmatrix} J' \\ M \end{pmatrix}, \tag{5.13}$$

wobei E^J durch (4.5) hinreichend genau bestimmt ist. Mit (5.6) und (5.10) erhalten wir daher in der gleichen Näherung

$$E_M^J = E^J + \mu_e g_J M B\,. \tag{5.14}$$

Hier ist $B \equiv B_z \equiv |\boldsymbol{B}|$. Der zweite Summand des in (5.9) angegebenen Ausdruckes für den Landéfaktor g_J rührt von der „ersten magnetischen Anomalie" des Elektrons her[*].

Der „anomalen" Zeemann-Aufspaltung (5.14), (5.9) entspricht eine „anomale" (und nur näherungsweise reguläre) Larmorpräzession. Ihre Frequenz $\boldsymbol{\omega}_{LS}$ ist das g_J-fache der „normalen" Larmorfrequenz $\boldsymbol{\omega}_L = -\bar{e}\,\boldsymbol{B}/2c\,m_e$ einer freien (d. h. nicht an die Spins angekoppelten) Bahnbewegung[**], wie sie etwa für s-Elektronen realisiert ist. Im Falle einer relativ starken Spin-Bahn-Kopplung ($\Omega^{LS} \gg \omega_L$) gilt nämlich

$$\dot{\boldsymbol{J}} = \hbar^{-1}\mu_e\,\boldsymbol{B} \times (\boldsymbol{L} + g_e\,\boldsymbol{S}) \approx \boldsymbol{\omega}_{LS} \times \boldsymbol{J}\,,$$

mit

$$\boldsymbol{\omega}_{LS} = g_J\,(e/2c\,m_e)\,\boldsymbol{B}$$

und

$$g_J = (\boldsymbol{L}_{\|} + g_e\,\boldsymbol{S}_{\|})/\sqrt{J(J+1)}\,,$$

in Übereinstimmung mit (5.9). Die um $\boldsymbol{J}$ mit der hohen Frequenz Ω^{LS} präzedierende Transversalkomponente

$$\boldsymbol{m}_\perp = -\mu_e(\boldsymbol{L}_\perp + g_e\,\boldsymbol{S})$$

---

[*] In der Atomhüllenphysik ist die Differenz $g_e - 1$, in der Elementarteilchenphysik hingegen die kleine Größe $\frac{1}{2}g_e - 1$ das „anomale magnetische Moment" des Elektrons.

[**] Freie Elektronenspins präzedieren natürlich ebenfalls regulär, und zwar mit der Frequenz $\boldsymbol{\omega}_S = g_e \cdot \boldsymbol{\omega}_L$.

des magnetischen Moments trägt in erster Näherung nichts zu der langsamen Larmorpräzession bei, da sich ihre Wirkung im Zeitmittel weghebt. Sie bewirkt aber eine gewisse Irregularität der Präzession, die sich in (5.4) und (5.5) an einem Unscharfwerden des Gesamtdrehimpulsbetrages (durch die Störterme mit $J' \neq J$) bemerkbar macht.

6. Das induzierte Magnetfeld

In großer Entfernung vom Atom erzeugt dessen Hülle ein Magnetfeld, das vollständig durch ihr in (5.1) angegebenes magnetisches Moment m bestimmt ist. Diese Größe hängt eng mit der Suszeptibilität zusammen; sie ist von VAN VLECK (1932) untersucht worden.

Uns hingegen interessiert hier das von der Hülle am Ort des Kerns erzeugte Magnetfeld b. Es lautet (ABRAGAM und PRYCE 1951)

$$b = \sum_{i=1}^{N} \left\{ \frac{\bar{e}}{c} \cdot \frac{r_i \times v_i}{r_i^3} + \mathrm{P} \frac{3\mu_i r_i \cdot r_i - r_i^2 \cdot \mu_i}{r_i^5} + \frac{8\pi}{3} \mu_i \, \delta(r_i) \right\}. \quad (6.1)$$

Der erste Summand bezeichnet das Feld der Elektronen-Bahnen gemäß dem Biot-Savartschen Gesetz; die beiden anderen Summanden beruhen auf den als punktförmig angenommenen Elektronen-Momenten μ_i. Dabei bedeutet das Hauptwertsymbol P, daß ein unendlich kleines Kügelchen um den singulären Punkt $r_i = 0$ als Mittelpunkt auszuschließen ist. Am Ort des Elektrons wird das Magnetfeld durch den dritten Summanden, den sog. Kontaktterm, gegeben [*].

In der Schalennäherung tragen abgeschlossene Schalen zu der Summe (6.1) nichts bei, und für die eine nicht abgeschlossene Schale (n, l) können die Variablen r_i^{-3} durch ihren nach (4.3) zu berechnenden Mittelwert $\langle r^{-3} \rangle$ ersetzt werden. Da im folgenden $l \neq 0$ (d. h. $l \neq 0$) vorausgesetzt ist, gilt stets $r_i \neq 0$. Folglich kann $\mathrm{P} = 1$ und $\delta(r_i) = 0$ gesetzt werden. Somit ist effektiv

$$b = -\mu_e \langle r^{-3} \rangle \sum_{k=1}^{q} \{ 2l_k + g_e (3s_k e_k \cdot e_k - s_k) \} - \gamma B. \quad (6.2)$$

Dabei ist $e_k \equiv r_k/r_k$ der die Richtung vom Kern zum k-ten Elektron anzeigende Einheitsvektor und

$$\gamma = \frac{e^2}{2c^2 m_e} \sum_{i=1}^{N} \frac{r_i^2 - r_i \cdot r_i}{r_i^3} \approx \frac{e^2}{3c^2 m_e} \sum_{i=1}^{N} \frac{1}{r_i}. \quad (6.3)$$

Die Summe $\sum r_i^{-1}$ ist von der Größenordnung $N(Z-z)/a_\mathrm{H}$, wobei z eine mittlere Abschirmungszahl bezeichnet; damit ist

$$\gamma \cong \frac{1}{3} N(Z-z)/137^2 \ll 1.$$

In vielen Fällen wird man daher die diamagnetische Korrektur vernachlässigen können. Das werden wir im folgenden der Einfachheit halber tun.

---

[*] FERMI (1930), HELMERS (1959).

Im Rahmen der Russell-Saunders-Näherung kann der Ausdruck (6.3) für das induzierte Magnetfeld weiter stark vereinfacht werden: Der erste Summand ist offenbar gleich $-\mu_e \langle r^{-3} \rangle \, 2\,L$. Der zweite muß aus Symmetriegründen (nach dem Wigner-Eckart-Theorem) proportional zu $3 \cdot \frac{1}{2}(SL \cdot L + L \cdot SL) - L^2 \cdot S$ sein; denn die Komponenten dieses Operators transformieren sich bei Drehungen der Orts- und der Spinvariablen genau so[*] wie die des Operators $\Sigma(r_k^2 \cdot s_k - 3r_k \cdot r_k\, s_k)$. Innerhalb eines gegebenen Multipletts ist somit

$$b = -\mu_e \langle r^{-3}\rangle \left\{ 2L + g_e \xi \left(L^2 \cdot S - \frac{3}{2}\, L \cdot LS + \frac{3}{2}\, SL \cdot L \right) - \gamma B \right\}. \quad (6.4)$$

Dies gilt für die verschiedenen Orientierungsmöglichkeiten der Vektoren L und S bei festen Beträgen L und S.

Der Koeffizient ξ hängt natürlich davon ab, mit welchem Multiplett man es zu tun hat. Für das durch die Hundschen Regeln angegebene Multiplett kann ξ (durch die Wahl einer besonders übersichtlichen Orientierung aller Bahn- und Eigendrehimpulse) verhältnismäßig leicht berechnet werden (vgl. Anhang 3). Das zuerst von Abragam und Pryce (1951) ohne Beweis angegebene Resultat lautet

$$\xi = \frac{2l + 1 - 4S}{(2l - 1)\,(2l + 3)\,(2L - 1)\,S}, \quad 0 < q < 4l + 2. \quad (6.5)$$

Hierin sind S und L nach (3.1) einzusetzen (mit $l = 3$ im Falle der Lanthaniden). Für die leere und volle Schale ($q = 0$ bzw. $\bar{q} = 0$) ist mit $S = L = 0$ aus Symmetriegründen auch $b = 0$, so daß ξ gar nicht definiert zu werden braucht.

Im folgenden benötigen wir lediglich die Erwartungswerte $\langle b \rangle_M^J$. Sie sind nach (5.5) zu berechnen, indem man für A der Reihe nach drei linear unabhängige Komponenten von b einsetzt. Aus Symmetriegründen verschwinden die auf B senkrechten Komponenten; es verbleibt allein die (mit der cartesischen Komponente b_z identische sphärische) Parallelkomponente b_0. Das wird durch die zu (5.6) analoge Wigner-Eckart-Formel

$$\begin{pmatrix} J \\ M \end{pmatrix} \Big| b_\mu \Big| \begin{pmatrix} J' \\ M \end{pmatrix} = \frac{(-1)^{J+1-J'}}{\sqrt{2J+1}} \begin{pmatrix} J & 1 & J' \\ M & \mu & M \end{pmatrix} (J \| b \| J') \quad (6.6)$$

bestätigt, der gemäß $|J' - 1| \leq J \leq J + 1$ und $M + \mu = M$ sein müssen.

Das reduzierte Matrixelement von b kann ähnlich dem von m ausgewertet werden, wenn man beachtet, daß innerhalb eines (L, S)-Multipletts $L^2 = L(L + 1)$ und $2LS = 2SL = J^2 - L(L + 1) - S(S + 1)$ ist, wobei J^2 durch $J(J + 1)$ oder durch $J'(J' + 1)$ ersetzt werden kann, je nachdem, ob dieser Operator auf den „linken" Vektor $(J|$ oder den „rechten" Vektor $|J')$ wirkt. So erhält man entsprechend zu (5.7) die

---

[*] Es handelt sich um das vektorielle Produkt eines Drehtensors zweiter Stufe im Ortsraum mit einem Vektor (d. h. einem Drehtensor erster Stufe) im Spinraum.

Beziehung

$$\begin{cases} \dfrac{(-1)^{L+S}(J\|\,\boldsymbol{b}\,\|J')}{\mu_e\langle r^{-3}\rangle\sqrt{(2J+1)\,(2J'+1)}} = 2(-1)^{-J'}\begin{Bmatrix} J & 1 & J' \\ L & S & L \end{Bmatrix}(L\|\,\boldsymbol{L}\,\|L) + \\[2mm] + g_e\,\xi\left\{(-1)^{-J}L\,(L+1)\begin{Bmatrix} J & 1 & J' \\ S & L & S \end{Bmatrix}(S\|\,\boldsymbol{S}\,\|S) - (-1)^{-J'}\times \\[2mm] \times\begin{Bmatrix} J & 1 & J' \\ L & S & L \end{Bmatrix}(L\|\,\boldsymbol{L}\,\|L)\dfrac{3}{4}\,[J\,(J+1)+J'\,(J'+1)-2L\,(L+1)-2S\,(S+1)]\right\}. \end{cases}$$

(6.7)

Einsetzen der Racah-Koeffizienten (nach EDMONDS 1957, S. 130) sowie der reduzierten Matrixelemente $(L\|\,\boldsymbol{L}\,\|L) = \sqrt{L\,(L+1)\,(2L+1)}$ und $(S\|\,\boldsymbol{S}\,\|S)$ $= \sqrt{S\,(S+1)\,(2S+1)}$, vgl. (5.8), führt unter Verwendung der zu (5.9) analogen Abkürzungen

$$\begin{cases} \gamma_J = 2\,[J\,(J+1)+L\,(L+1)-S\,(S+1)]+g_e\,\xi\left\{L\,(L+1)\cdot[J\,(J+1)- \right. \\[2mm] \quad -L\,(L+1)+S\,(S+1)] - \dfrac{3}{2}\,[J\,(J+1)+L\,(L+1)-S\,(S+1)]\times \\[2mm] \quad \left.\times[J\,(J+1)-L\,(L+1)-S\,(S+1)]\right\}, \\[2mm] \eta_J = 2-g_e\,\xi\cdot\dfrac{1}{2}\,[3J^2-L\,(L+1)-3S\,(S+1)] \end{cases}$$

(6.8)

zu den reduzierten Feldstärken

$$\begin{cases} (J\|\,\boldsymbol{b}\,\|J) = -\mu_e\langle r^{-3}\rangle\dfrac{1}{2}\sqrt{\dfrac{2J+1}{J\,(J+1)}}\cdot\gamma_J, \\[3mm] (J\|\,\boldsymbol{b}\,\|J-1) = -(J-1\|\,\boldsymbol{b}\,\|J) = -\mu_e\langle r^{-3}\rangle\dfrac{1}{2}\sqrt{h_J}\cdot\eta_J. \end{cases}$$

(6.9)

Für $|J'-J| \geqq 2$ und $J = J' = 0$ ist $(J\|\,\boldsymbol{b}\,\|J') = (J'\|\,\boldsymbol{b}\,\|J) = 0$.

7. Die thermische Mittelung

Die gesuchte Größe ist der thermische Mittelwert $\langle\boldsymbol{b}\rangle$ des induzierten Feldes. Nach BOLTZMANN und GIBBS hängt er mit den quantenmechanischen Mittelwerten $\langle\boldsymbol{b}\rangle_M^J$ der thermisch angeregten Zustände folgendermaßen zusammen

$$\langle\boldsymbol{b}\rangle\cdot\sum_J\sum_M e^{-E_M^J/kT} = \sum_J\sum_M \langle\boldsymbol{b}\rangle_M^J\, e^{-E_M^J/kT}.$$

(7.1)

Hier ist vorausgesetzt, daß nur das Grundmultiplett mit merklicher Wahrscheinlichkeit besetzt wird. Das wird genau dann der Fall sein, wenn die Anregungsenergie der höheren Multipletts groß gegen kT ist. Eine notwendige und bis auf zufällige Ausnahmen auch hinreichende Bedingung dafür ist

$$kT \ll e^2/a_H.$$

Über das Verhältnis der thermischen Energie kT zur Spin-Bahn-Energie $\hbar \Omega^{LS} [= \eta \zeta \sqrt{J(J+1)}]$ braucht nichts vorausgesetzt zu werden.

Zunächst mitteln wir über alle $2J+1$-Zustände eines Zeemanaufgespaltenen Niveaus, berechnen also das von dem J-Niveau erzeugte Feld $\langle b \rangle^J$. Es ist durch die Gleichung

$$\langle b \rangle^J \sum_M e^{-E_M^J/kT} = \sum_M \langle b \rangle_M^J e^{-E_M^J/kT} \tag{7.2}$$

gegeben, wobei E_M^J nach (5.14) und $\langle b \rangle_M^J$ nach (5.5), (5.6), (6.6), einzusetzen ist. Da Absättigungseffekte von uns vernachlässigt werden, das angelegte Feld B also nur in erster Näherung berücksichtigt wird, kann die Vereinfachung

$$e^{-E_M^J/kT} = e^{-E^J/kT} \cdot (1 - \mu_e g_J M B/kT)$$

vorgenommen werden. Dabei wird vorausgesetzt, daß

$$J \mu_e B \ll kT$$

ist*.

Der Beitrag nullter Ordnung verschwindet aus Symmetriegründen ($\Sigma M = 0$); d. h. ohne äußeres Feld gibt es im Mittel über alle Orientierungen der Ionen auch kein induziertes paramagnetisches Feld. Mit dem äußeren Feld aber wird

$$\sum_M e^{-E_M^J/kT} = e^{-E^J/kT} \cdot (2J+1)$$

und

$$\Sigma \langle b \rangle_M^J \cdot e^{-E_M^J/kT}$$

$$= e^{-E^J/kT} \sum_{J'} \left(\frac{\delta_{J'J}}{kT} + 2 \frac{1-\delta_{J'J}}{E^{J'}-E^J} \right) \frac{(-1)^{J'-J}}{3} (J' \| m \| J) \, (J \| b \| J') \, B. \tag{7.3}$$

Hierbei ist folgende in den Symmetrie- und Orthogonalitätseigenschaften der Clebsch-Gordan-Koeffizienten** enthaltene Aussage benutzt worden

$$\frac{3(-1)^{J-J'}}{\sqrt{(2J+1)(2J'+1)}} \sum_M \begin{pmatrix} J' & 1 & J \\ M & 0 & M \end{pmatrix} \begin{pmatrix} J & 1 & J' \\ M & 0 & M \end{pmatrix} = \sum_M \begin{pmatrix} J' & J & 1 \\ -M & M & 0 \end{pmatrix}^2 = 1, \tag{7.4}$$

sofern das Tripel (J', J, 1) die (in seinen drei Elementen symmetrische) Dreieckrelation $|J'-J| \leq 1 \leq J'+J$ erfüllt. Anderenfalls würde man 0 statt 1 erhalten; doch tragen die betreffenden Terme ohnehin nichts zu der Summe in (7.3) bei, da ihre reduzierten Matrixelemente verschwinden.

Nun mitteln wir über sämtliche Niveaus des Grundmultipletts. Einsetzen von (4.5), (5.10) und (6.9) in (7.3) und hiermit in (7.4) liefert

---

* Für $T = 300$ Grad und $J = 4$ muß danach $B \ll 10^6$ Gs sein.

** Diese denken wir uns wie üblich als reell gewählt (die Quantisierungsachsen aller Vektoren sollen zusammenfallen).

das Resultat

$$\frac{\langle b \rangle}{B} = \frac{\mu_e^2 \langle r^{-3} \rangle}{6} \times$$

$$\times \frac{\sum_J \left[(2J+1)\frac{g_J \gamma_J}{kT} + \frac{g_e - 1}{\frac{1}{2}\zeta\eta} \cdot \left(\frac{h_J \eta_J}{J} - \frac{h_{J+1}\eta_{J+1}}{J+1} \right) \right] e^{-\frac{1}{2}\eta\zeta J(J+1)/kT}}{\sum_J (2J+1)\, e^{-\frac{1}{2}\eta\zeta J(J+1)/kT}} .$$

(7.5)

Summiert wird von $|L-S|$ bis $L+S$. Dabei ist $h_{|L-S|} = \eta_{|L-S|} = 0$ und $h_{L+S+1} = \eta_{L+S+1} = 0$, so daß man (7.5) auch folgendermaßen umformen kann

$$\frac{\langle b \rangle}{B} = \frac{\mu_e^2 \langle r^{-3} \rangle}{6\,kT} \times$$

$$\times \frac{\sum_J \left[(2J+1)\, g_J \gamma_J + (g_e - 1)\frac{h_J \eta_J (1 - e^{2\nu J})}{J y} \right] e^{-\nu J(J+1)}}{\sum_J (2J+1)\, e^{-\nu J(J+1)}} .$$

(7.6)

Dazu ist der dimensionslose Parameter

$$y = \frac{\eta\,\zeta}{2\,kT}$$

(7.7)

eingeführt worden. Die Koeffizienten η, g_J, γ_J, h_J, η_J kann man den Formeln (4.4), (5.9), (6.8) entnehmen, während ζ in (4.2) angegeben ist.

Ein schwierigeres Problem stellt der nach (3.3) zu berechnende Koeffizient $\langle r^{-3} \rangle$ dar. Sein Reziprokes ist ein Maß für das „effektive Volumen" der $4f$-Schale in bezug auf den Paramagnetismus am Kernort.

Die Größenordnung des Parameters y liegt zufällig in der Umgebung von 1, wenn T von der Größenordnung der Zimmertemperatur ist. Falls für das Grundniveau dennoch $e^{2J\nu} \gg 1$ gesetzt werden darf, vereinfacht sich die Endformel beträchtlich, denn man braucht die thermischen Anregungen nicht zu berücksichtigen. Ist sogar die Ungleichung $Jy \gg 1$ hinreichend gut erfüllt, so können auch die vom äußeren Magnetfeld induzierten „virtuellen Anregungen" der ungestörten Niveaus vernachlässigt werden. Man erhält so die zuerst von GOLDRING und SCHARENBERG (1958) angegebene vereinfachte Formel

$$\frac{\langle b \rangle}{B} = \frac{\mu_e^2 \langle r^{-3} \rangle}{6\,kT}\, \gamma_J g_J ,$$

(7.8)

die folgende Interpretation zuläßt

$$\langle b_z \rangle = \langle b \rangle \cdot \frac{\langle J_z \rangle}{J} ,$$

(7.9)

wobei $\langle b \rangle$ die Feldstärke am Kernort und $\langle J_z \rangle$ das thermische Mittel des Drehimpulses in Magnetfeldrichtung sind. Und zwar gilt[*]

$$\langle b \rangle \equiv \left(\frac{J}{J} \bigg| b_z \bigg| \frac{J}{J} \right) = -\frac{1}{2}\,\mu_e \langle r^{-3} \rangle\, \gamma_J / (J+1)$$

(7.10)

---

[*] Mit $\langle r^{-3} \rangle = \text{Å}^{-3}$, $J = 4$ und $\gamma_J = 100$ erhält man den Schätzwert $\langle b \rangle \approx \approx 10^5$ Gs.

und

$$\langle J_z \rangle \equiv \frac{\sum\limits_{M} M \cdot e^{-\mu_e g_J \cdot MB/kT}}{\sum\limits_{M} e^{-\mu_e g_J MB/kT}} = \frac{\mu_e g_J J(J+1) B}{3kT} \qquad (7.11)$$

wegen $\Sigma M^2 = \frac{1}{3} J(J+1)(2J+1)$. Der Drehimpuls des Grundzustandes ergibt sich aus (3.1) und dem Vorzeichen von (4.4) zu

$$J = \begin{cases} \dfrac{1}{2}\, q(2l - q) & \text{für} \quad 0 \leqq q \leqq 2l \\[2mm] \dfrac{1}{2}\, \bar{q}(2l + 2 - \bar{q}) & \text{für} \quad 2l + 2 \geqq \bar{q} \geqq 0. \end{cases} \qquad (7.12)$$

Für $q = 2l + 1 = \bar{q}$ (also Gd^{+++}) ist $L = 0$; folglich muß das induzierte Magnetfeld am Ort des Kerns verschwinden, da die Zustandsfunktion im Ortsraum kugelsymmetrisch zum Kern ist. Nur für Pm^{+++}, Sm^{+++} und Eu^{+++} ist J_y nicht groß genug, die L-S-Präzession also nicht schnell genug, um die anomalen Schwankungen der Larmorpräzession vernachlässigbar klein zu machen.

8. Korrekturen

Der spektroskopische Vergleich der betrachteten Multiplettstruktur mit der bei reiner Russell-Saunders-Kopplung zu erwartenden zeigt, daß die LS-Kopplung die tatsächlichen Verhältnisse recht gut beschreibt, daß aber systematisch kleine Abweichungen auftreten, die auf eine gewisse Lockerung dieser Kopplung hinweisen, so daß L und S keine exakten Quantenzahlen mehr sind. Glücklicherweise sind jedoch die Abweichungen klein genug, um durch Störungsrechnung zweiter Ordnung erfaßt werden zu können. Als Störenergie ist der Differenzoperator

$$\Delta = \zeta \left\{ \sum_{k=1}^{q} l_k s_k - \eta\, LS \right\} \qquad (8.1)$$

einzusetzen. Die für die Störung maßgebenden Matrixelemente[*] $\left(\begin{smallmatrix} (L', S')\, J' \\ M' \end{smallmatrix} \middle| \Delta \middle| \begin{smallmatrix} (L, S)\, J \\ M \end{smallmatrix} \right)$ sind proportional zu den Kronecker-Koeffizienten $\delta_{JJ'}$ und $\delta_{MM'}$ und zu dem Racah-Koeffizienten $\left\{ \begin{smallmatrix} J & L & S \\ 1 & S' & L' \end{smallmatrix} \right\}$.

Fast ebenso wichtig wie diese Korrektur ist die Berücksichtigung der im 5. Abschnitt vernachlässigten diamagnetischen Terme und der dort ganz außer Betracht gebliebenen relativistischen Effekte. Hierzu hat man auszugehen von der Diracschen Gleichung für ein System von

---

[*] Unter $\left| \begin{smallmatrix} (L, S)\, J \\ M \end{smallmatrix} \right)$ verstehen wir nichts anderes als die Wellenfunktion $\left| \begin{smallmatrix} J \\ M \end{smallmatrix} \right)$ von (4.6). Dort konnten wir die Quantenzahlen L, S weglassen, da das Multiplett festgehalten wurde.

Elektronen in einem äußeren elektro- und magnetostatischen Feld mit Breitschen Wechselwirkungstermen für die Elektronenpaare (BREIT 1928, 1929; MARGENAU 1940). Mit Hilfe einer Foldy-Wouythuysen-Transformation (FOLDY und WOUYTHUYSEN 1950) hat man dies auf die Form einer Schrödinger-Pauli-Gleichung zu bringen. Schließlich sind die so erhaltenen Zusatzterme in zweiter störungstheoretischer Ordnung auszuwerten.

Alle bisher genannten Korrekturen wurden von JUDD und LINDGREN (1961) für die durch die Gleichung

$$\left(\begin{matrix} (L, S) \ J \\ M \end{matrix} \middle| \boldsymbol{L} + g_e \boldsymbol{S} \middle| \begin{matrix} (L, S) \ J \\ M \end{matrix}\right) = M \, g_{(L, S)\, J} \tag{8.2}$$

definierten Landéschen Faktoren $g_{(L, S)\, J}$ der Grundmultipletts sämtlicher Lanthaniden berechnet und tabelliert. Wo immer diese g_J-Werte benötigt werden, empfiehlt es sich, sie nicht nach der bekannten Formel (5.9) zu bestimmen, sondern die Tabellenwerte von JUDD und LINDGREN zu verwenden. Sehr zu begrüßen wäre eine Berechnung der entsprechenden Korrekturen zu den Gln. (5.9) und (6.8) für die Koeffizienten $h_{(L, S)\, J}$, $\gamma_{(L, S)\, J}$ und $\eta_{(L, S)\, J}$. Bislang sind uns solche Rechnungen leider nicht bekannt geworden.

Korrekturen ganz anderer Art betreffen die in (4.2) und (7.8) eingeführten Mittelwerte $\langle [Z - z(r)]\, r^{-3} \rangle$ und $\langle r^{-3} \rangle$. Die Mittelung nach (4.3) verwendet die üblichen Hartreeschen Gleichungen. Wir haben sie unseren Formeln zugrundegelegt, um den physikalischen Gedankengang von Komplikationen frei zu halten, die mit der Struktur des hier behandelten Paramagnetismus nichts zu tun haben. Will man genauere Werte von $\langle [Z - z(r)]\, r^{-3} \rangle$ und $\langle r^{-3} \rangle$ berechnen, so muß man ein genaueres Modell der Atomhülle zugrundelegen. Das hat aber auf die weiteren Überlegungen keinen Einfluß, solange keine Konfigurationsmischungen zugelassen werden, d. h. solange n, l (und damit auch q) gute Quantenzahlen bleiben. Verbesserungen der numerischen Werte sind im Rahmen der hier entwickelten Theorie in drei verschiedenen Richtungen möglich, entsprechend den drei Hartreeschen Näherungsannahmen: (i) Forderung der Kugelsymmetrie, (ii) Vernachlässigung der Austauscheffekte, (iii) Verzicht auf Diagonalisierung des Drehimpulsbetrages bei der Bestimmung des selbstkonsistenten* Feldes. Daß die Annahme (i) nicht selbstverständlich ist, zeigen im analogen Fall des Kernschalenmodells die stark deformierten Atomkerne. Zur Rechtfertigung der Annahme für die Atomhülle kann aber folgendes gesagt werden: Da das elektrostatische Feld des Atomkerns, durch welches die Atomhülle zusammengehalten wird, Zentralsymmetrie besitzt, ist das kugelsymmetrische Schalenmodell zum mindesten eine

---

* Durch diese drei Annahmen wird die Selbstkonsistenz ein wenig verletzt, denn durch die erhaltenen Wellenfunktionen (die ja nachträglich antimetrisiert und zu scharfem J kombiniert werden) werden die vereinfachenden Forderungen offenbar nur teilweise erfüllt.

brauchbare Ausgangsnäherung*. Die Ladungsdichte der abgeschlossenen Schalen ist jedoch streng kugelsymmetrisch. Das folgt mit gruppentheoretischen Mitteln aus der Tatsache, daß deren Gesamtbahndrehimpuls L verschwindet, oder auch im Rahmen des Schalenmodells elementar aus der Unsöldschen Identität

$$\sum_{m=-l}^{l} |Y_{lm}(\vartheta, \varphi)|^2 = \frac{2l+1}{4\pi}. \qquad (8.3)$$

Sie ist ein Ausdruck des Sachverhalts, daß jede Drehung die $2l+1$ Vektoren $Y_{ll}, \ldots, Y_{l,-l}$ einer unitären Transformation unterwirft. Eine kugelsymmetrische Ladung erzeugt aber ein zentralsymmetrisches Feld. Die Zentralsymmetrie kann mithin höchstens durch die wenigen Leuchtelektronen gestört werden, und die Näherung (i) besteht lediglich darin, daß für diese die Richtungsverteilungen $|Y_{lm}(\vartheta, \varphi)|^2$ durch ihre isotropen Mittelwerte $(4\pi)^{-1}$ ersetzt werden. Für das Feld ist der begangene Fehler noch bedeutend geringer (besonders für große l), da es aus der Ladungsdichte durch zweifache Integration hervorgeht, wodurch Inhomogenitäten und Anisotropien geglättet werden. Zu diesen geometrischen Tatsachen kommt noch die dynamische Tendenz zu einer kugelsymmetrischen Ausbildung der Ladungsverteilung hinzu, die darauf beruht, daß die elektrische Abstoßung der Elektronen im zentralsymmetrischen Kernfeld eine gleichmäßige Raumausfüllung und somit auch Kugelsymmetrie bevorzugt**. Annahme (ii) bedeutet den Übergang von den Fockschen (Fock 1930) zu den sehr viel einfacheren Hartreeschen Gleichungen. Annahme (iii) stellt eine analoge Vernachlässigung dar, die in der Literatur meist stillschweigend gemacht wird und noch kaum untersucht ist.

Falls Konfigurationsmischungen mit ungepaarten Elektronen zugelassen werden, darf der Kontaktterm in (6.1) nicht übersehen werden.

Anhang 1: Die ungestörten Wellenfunktionen

Nach der Hundschen Regel haben die ungestörten Wellenfunktionen $\left|\begin{smallmatrix} L & S \\ \Lambda & \Sigma \end{smallmatrix}\right)$ eine einfache Symmetrie bei Vertauschung der Elektronen. Die Funktionen werden besonders einfach, wenn die z-Komponenten Λ und Σ (von L bzw. S) ihre maximalen Werte L bzw. S annehmen. Wir teilen sie nacheinander für die beiden Fälle $q \leq 2l+1$ und $q \geq 2l+1$ mit.

---

* Hierin unterscheidet sich das Schalenmodell der Atomhülle wesentlich von dem des Atomkerns: Für die stark ellipsoidischen Kerne muß man von der Nilssonschen Verallgemeinerung des Mayer-Jensenschen Modells ausgehen; vgl. M. G. Mayer und J. H. D. Jensen: "Elementary Theory of Nuclear Shell Structure" (1955), S. G. Nilsson, Mat. Fys. Medd. 29, no. 16 (1955) und K. Gottfried, Phys. Rev. 103, 1017 (1956).

** Auch hierin sind die Verhältnisse im Atomkern fast entgegengesetzt denen in der Atomhülle: Die attraktiven Kernkräfte endlicher Reichweite enthalten eine starke Tendenz zur Deformation der Atomkerne, die sich in vielen Fällen gegen die ebenfalls vorhandene Neigung zur Kugelgestalt ein Stück weit durchsetzen kann.

I. Teilchenkonfigurationen $(q \leqq 2l + 1)$, d. h. die Unterschale der Leuchtelektronen ist höchstens halb voll. Hier ist

$$\begin{vmatrix} L & S \\ L & S \end{vmatrix} = \begin{vmatrix} L \\ L \end{vmatrix} \begin{vmatrix} S \\ S \end{vmatrix} \tag{A1.1}$$

mit*

$$\begin{vmatrix} L \\ L \end{vmatrix} = \sqrt{q!} \;\; \mathscr{A}_q \begin{vmatrix} l_1 \\ l \end{vmatrix} \begin{vmatrix} l_2 \\ l - 1 \end{vmatrix} \cdots \begin{vmatrix} l_q \\ l - q + 1 \end{vmatrix} \tag{A1.2}$$

und*

$$\begin{vmatrix} S \\ S \end{vmatrix} = \begin{vmatrix} s_1 \\ 1/2 \end{vmatrix} \begin{vmatrix} s_2 \\ 1/2 \end{vmatrix} \cdots \begin{vmatrix} s_q \\ 1/2 \end{vmatrix} . \tag{A1.3}$$

Mit $\mathscr{A}_q$ bezeichnen wir den Projektionsoperator der Antimetrisierung q-stelliger Funktion**.

Sämtliche Bahnfunktionen $\begin{vmatrix} L \\ L \end{vmatrix}$ sind hier voll antimetrisch, haben also den Symmetriecharakter $[1^q]$. Die Spinfunktionen $\begin{vmatrix} S \\ S \end{vmatrix}$ hingegen sind hier voll symmetrisch, gehören also zum Charakter $[q]$. Allgemein bezeichnet $[z_1^{\alpha_1}, z_2^{\alpha_2}, \ldots, z_k^{\alpha_k}]$ ein Youngsches Diagramm*** mit α_1 Zeilen der Länge z_1, α_2 Zeilen der Länge z_2 usw., wobei $z_1 > z_2 > \cdots > z_k$ und $\alpha_1 z_1 + \alpha_2 z_2 + \cdots + \alpha_k z_k = q$ ist.

Die Ortsfunktion (A1.2) ist, von dem Normierungsfaktor $(q!)^{-1/2}$ abgesehen, eine Determinante. Ebenso kann natürlich auch (A1.1) als eine Determinante verstanden werden: man braucht nur die Spalten von (A1.2) mit den entsprechenden Faktoren von (A1.3) zu multiplizieren. So erhält man die $(q!)^{-1/2}$-fache Slaterdeterminante

$$\begin{vmatrix} L & S \\ L & S \end{vmatrix} = \sqrt{q!} \;\; \mathscr{A}_q \begin{vmatrix} l_1 & s_1 \\ l & \frac{1}{2} \end{vmatrix} \begin{vmatrix} l_2 & s_2 \\ l - 1 & \frac{1}{2} \end{vmatrix} \cdots \begin{vmatrix} l_q & s_q \\ l - q + 1 & \frac{1}{2} \end{vmatrix} \tag{A1.4}$$

Sie hat selbstverständlich den Symmetriecharakter $[1^q]$.

II. Löcherkonfiguration $(q \geqq 2l + 1)$, d. h. die Unterschale der Leuchtelektronen ist mindestens halb voll. Diesmal haben wir die Slaterdeterminante

$$\begin{vmatrix} L & S \\ L & S \end{vmatrix} = \sqrt{q!} \;\; \mathscr{A}_q \begin{vmatrix} l_1 & s_1 \\ l & \frac{1}{2} \end{vmatrix} \cdots \begin{vmatrix} l_{2l+1} & s_{2l+1} \\ -l & \frac{1}{2} \end{vmatrix} \times$$
$$\times \begin{vmatrix} l_{2l+2} & s_{2l+2} \\ l & -\frac{1}{2} \end{vmatrix} \cdots \begin{vmatrix} l_q & s_q \\ q - l & -\frac{1}{2} \end{vmatrix} , \tag{A1.5}$$

die sich nicht in eine Orts- und eine Spin-Funktion faktorisieren läßt. Doch läßt sie sich in eine Summe von Produkten von Bahn- und Spin-Funktionen mit verschiedenen Teilchenzuordnungen schreiben. (Zu

---

* Die Indizes an l und s bezeichnen die Nummer des betreffenden Elektrons. Sämtliche l_j bzw. s_j haben also den gleichen festen Wert l bzw. s (für die Lanthaniden, also 3 bzw. $1/2$).

** Explizit: $\mathscr{A}_q f(1, \ldots, q) = (q!)^{-1} \Sigma \varepsilon_P f(P1, \ldots, Pq)$, wobei P alle $q!$ Permutationen durchläuft.

*** Siehe z. B. M. HAMMERMESH: "Group Theory and its Application to Physical Problems" (London 1962) oder D. E. RUTHERFORD: "Substitutional Analysis" (Edinburgh 1948).

einem Youngschen Diagramm gehören im allgemeinen mehrere Youngsche Tableaux, die auseinander durch Permutationen der Teilchen hervor-
gehen.) Dabei haben die Bahnfunktionen $\begin{vmatrix} L \\ L \end{vmatrix}$ die Symmetrie $[2^{2l+1-\bar{q}}, 1^{\bar{q}}]$
und die Spinfunktion $\begin{vmatrix} S \\ S \end{vmatrix}$ die Symmetrie $[2l+1, 2l+1-\bar{q}]$ gegenüber
Vertauschungen von Teilchen (und nicht etwa von Löchern).

Die Determinante (A1.5) stellt ebenso wie (A1.4) eine Unter-
determinante der Slaterdeterminante der vollen l-Schale dar, welche die
Austauschsymmetrie $[1^{4l+2}]$ besitzt und in jeder Hinsicht sphärisch-
symmetrisch ist ($L = S = J = 0$). Die beiden Unterdeterminanten sind
zueinander adjungiert. Hierauf beruht nach dem Laplaceschen Ent-
wicklungssatz die Symmetrie gegenüber Vertauschung von Teilchen mit
Löchern, bei der q in $\bar{q} = 4l + 2 - q$ übergeht. Diese Algebra wird mit
Hilfe des feldtheoretischen Formalismus erheblich vereinfacht.

Anhang 2: Berechnung des Spin-Bahn-Koeffizienten η

Nach dem Wigner-Eckart-Theorem sind die Matrixelemente von
$\sum l_k s_k$ und LS proportional in dem Sinne, daß ihr Quotient

$$\eta \equiv \frac{\begin{pmatrix} L & S \\ \Lambda' & \Sigma' \end{pmatrix} \sum\limits_{k=1}^{q} l_k s_k \begin{vmatrix} L & S \\ \Lambda & \Sigma \end{vmatrix}}{\begin{pmatrix} L & S \\ \Lambda' & \Sigma' \end{pmatrix} LS \begin{vmatrix} L & S \\ \Lambda & \Sigma \end{vmatrix}} \tag{A2.1}$$

wohl von den Drehimpulsbeträgen L und S abhängt, nicht jedoch von
den Orientierungsquantenzahlen Λ, $\Lambda'(=L_z \equiv L_0)$ und Σ, $\Sigma'(=S_z \equiv S_0)$.
Daher können wir η im *einfachsten Fall* auswerten, nämlich für $\Lambda = \Lambda'$
$= L$ und $\Sigma = \Sigma' = S$. Dabei ist die Fallunterscheidung des Anhanges 1
vorzunehmen, abgesehen von dem trivialen Fall $q = 2l + 1$.

I. Teilchenkonfigurationen ($q < 2l + 1$). Da die Wellenfunktionen
gegenüber gleichzeitigen Vertauschungen der Orts- und Spin-Koordinaten
antimetrisch sind, kann in (A2.1) jeder Summand $l_k s_k$ durch einen
beliebigen anderen ersetzt werden. Statt $\sum l_k s_k$ können wir daher $q\,ls$
schreiben, wobei es gleichgültig ist, welchen Index k die beiden Vektoren l und s hier tragen (so daß wir ihn gar nicht hinzuschreiben
brauchen). Danach ist

$$\eta = \frac{q \cdot \begin{pmatrix} L & S \\ L & S \end{pmatrix} ls \begin{vmatrix} L & S \\ L & S \end{vmatrix}}{\begin{pmatrix} L & S \\ L & S \end{pmatrix} LS \begin{vmatrix} L & S \\ L & S \end{vmatrix}} . \tag{A2.2}$$

Nach Anhang 1 ist somit

$$\begin{pmatrix} L & S \\ L & S \end{pmatrix} LS \begin{vmatrix} L & S \\ L & S \end{pmatrix} = \begin{pmatrix} L \\ L \end{pmatrix} L \begin{vmatrix} L \\ L \end{pmatrix} \begin{pmatrix} S \\ S \end{pmatrix} S \begin{vmatrix} S \\ S \end{pmatrix}$$

$$= \begin{pmatrix} L \\ L \end{pmatrix} L_0 \begin{vmatrix} L \\ L \end{pmatrix} \begin{pmatrix} S \\ S \end{pmatrix} S_0 \begin{vmatrix} S \\ S \end{pmatrix} = L \cdot S \tag{A2.3}$$

und

$$\left(\begin{matrix} L & S \\ L & S \end{matrix}\middle| \boldsymbol{l s}\middle| \begin{matrix} L & S \\ L & S \end{matrix}\right) = \left(\begin{matrix} L \\ L \end{matrix}\middle| \boldsymbol{L}\middle| \begin{matrix} L \\ L \end{matrix}\right)\left(\begin{matrix} S \\ S \end{matrix}\middle| \boldsymbol{S}\middle| \begin{matrix} S \\ S \end{matrix}\right)$$

$$= \left(\begin{matrix} L \\ L \end{matrix}\middle| l_0\middle| \begin{matrix} L \\ L \end{matrix}\right)\left(\begin{matrix} S \\ S \end{matrix}\middle| s_0\middle| \begin{matrix} S \\ S \end{matrix}\right) = \frac{L}{q}\cdot\frac{1}{2}\,. \tag{A2.4}$$

Dabei wurde benutzt, daß $S_{\pm 1}$ und $s_{\pm 1}$ die z-Komponente von $\boldsymbol{S}$ um 1 erhöhen bzw. erniedrigen (so daß die diagonalen Matrixelemente dieser Operatoren verschwinden müssen); und daß aus Symmetriegründen

$$q\left(\begin{matrix} L \\ L \end{matrix}\middle| \boldsymbol{l}\middle| \begin{matrix} L \\ L \end{matrix}\right) = \sum_{k=1}^{q}\left(\begin{matrix} L \\ L \end{matrix}\middle| \boldsymbol{l}_k\middle| \begin{matrix} L \\ L \end{matrix}\right) \tag{A2.5}$$

ist, also

$$q\left(\begin{matrix} L \\ L \end{matrix}\middle| l_0\middle| \begin{matrix} L \\ L \end{matrix}\right) = \sum_{k=1}^{q}(l-q+k) \equiv \sum_{l-q+1}^{\lambda=l}\lambda = \frac{q}{2}(2l-q+1) = L\,, \tag{A2.6}$$

vgl. (3.1). Hierin hat λ die Bedeutung von l_z. Da offenbar $s_0 = {}^1/_2$ ist, folgt aus (A2.2), (A2.3), (A2.4) und (3.1) die Behauptung $\eta = 1/2S = 1/q$ von (4.4).

II. Löcherkonfigurationen ($q > 2l + 1$). Diesen Fall kann man auf den vorherigen zurückführen: Man ersetze die Summe über die Teilchen durch die negative Summe über die Löcher, wobei alle Vektoren $\boldsymbol{l}$ und $\boldsymbol{s}$ ihre Vorzeichen wechseln. Da der wirkliche Zustand dadurch nicht geändert wird, bleiben $\boldsymbol{L}$ und $\boldsymbol{S}$ invariant. Folglich wird $\eta = -1/\bar{q}$ $= -1/2S$, wieder in Übereinstimmung mit (4.4) und (3.1). Auf eine ausführliche Begründung dieser Teilchen-Löcher-Transformation können wir verzichten. Sie ergibt sich ohne Schwierigkeit durch Vergleich des „Teilchenzustandes" (A1.1) mit dem entsprechenden „Löcherzustand" (A1.4).

Anhang 3: Berechnung des Spinfeld-Koeffizienten $\boldsymbol{\xi}$

Die Argumentation ist genau analog der des vorhergehenden Anhanges, nur sind die Formeln ein wenig komplizierter. Zu berechnen ist der nur von L und S abhängige Quotient

$$\xi = \frac{\left(\begin{matrix} L & S \\ \Lambda' & \Sigma' \end{matrix}\middle| U_\mu\middle| \begin{matrix} L & S \\ \Lambda & \Sigma \end{matrix}\right)}{\left(\begin{matrix} L & S \\ \Lambda' & \Sigma' \end{matrix}\middle| V_\mu\middle| \begin{matrix} L & S \\ \Lambda & \Sigma \end{matrix}\right)} \tag{A3.1}$$

in dem die Vektoren $\boldsymbol{U}$ und $\boldsymbol{V}$ folgende Bedeutung haben

$$\boldsymbol{U} = \sum_{k=1}^{q}\boldsymbol{u}_k \quad \text{mit} \quad \boldsymbol{u} = \frac{3\boldsymbol{r}\cdot\boldsymbol{r s} - r^2\cdot\boldsymbol{s}}{r^2} \tag{A3.2}$$

und

$$\boldsymbol{V} = 3\cdot\frac{\boldsymbol{L}\cdot\boldsymbol{L S} + \boldsymbol{S L}\cdot\boldsymbol{L}}{2} - L^2\cdot\boldsymbol{S}\,. \tag{A3.3}$$

Dieses $\boldsymbol{u}$ ist ein Einelektronenoperator und trägt daher einen von 1 bis q laufenden Index k, der die Elektronen der Unterschale durchnumeriert. Der Index μ unterscheidet stets die Vektorkomponenten; und zwar ist μ entweder $=x, y, z$ (cartesische Komponenten) oder $= +1, 0, -1$ (sphärische Komponenten), wobei stets $A_0 \equiv A_z$ und $A_{\pm 1} \equiv \mp (A_x \pm i A_y)/\sqrt{2}$ ist. Wir führen den Einheitsvektor $\boldsymbol{e} \equiv \boldsymbol{r}/r$ ein. Seine sphärischen Komponenten sind bekanntlich bis auf einen Normierungsfaktor identisch mit den drei Kugelflächenfunktionen erster Ordnung: $e_\mu \equiv \sqrt{4\pi/3}\ Y_{1\mu}(\boldsymbol{e})$. Wieder sind zwei Fälle zu unterscheiden.

I. Teilchenkonfigurationen. Aus Symmetriegründen ist

$$\xi = q \cdot \frac{\begin{pmatrix} L & S \\ L & S \end{pmatrix} u_0 \begin{vmatrix} L & S \\ L & S \end{vmatrix}}{\begin{pmatrix} L & S \\ L & S \end{pmatrix} V_0 \begin{vmatrix} L & S \\ L & S \end{vmatrix}} \ . \tag{A3.4}$$

Hierin kann $u_0 = 3e_0\,\boldsymbol{e}\boldsymbol{s} - s_0$ durch $(3e_0^2 - 1)\,s_0$ ersetzt werden, da ja $s_{\pm 1}$ jeden Spinzustand in einen orthogonalen überführt. Aus einem genau entsprechenden Grunde kann V_0 zu $(3L_0^2 - \boldsymbol{L}^2)\,S_0$ vereinfacht werden. Demnach ist

$$\begin{pmatrix} L & S \\ L & S \end{pmatrix} V_0 \begin{vmatrix} L & S \\ L & S \end{vmatrix} = [3L^2 - L(L+1)]\,S = (2L-1)\,L \cdot S \ ,$$

also

$$\xi = q \cdot \frac{1}{q} \cdot \frac{\overset{\lambda = l}{\underset{l-q+1}{\sum}} \begin{pmatrix} l \\ \lambda \end{pmatrix} 3e_0^2 - 1 \begin{vmatrix} l \\ \lambda \end{vmatrix} \cdot \frac{1}{2}}{(2L-1)\,LS} \ . \tag{A3.5}$$

Das verbliebene Matrixelement ist elementar auswertbar, zweckmäßigerweise wieder mit Hilfe des Wigner-Eckart-Theorems[*]

$$\begin{pmatrix} l \\ \lambda \end{pmatrix} 3e_0^2 - 1 \begin{vmatrix} l \\ \lambda \end{vmatrix} = \frac{\begin{pmatrix} l \\ \lambda \end{pmatrix} 3l_0^2 - 1 \begin{vmatrix} l \\ \lambda \end{vmatrix}}{\begin{pmatrix} l \\ l \end{pmatrix} 3l_0^2 - 1 \begin{vmatrix} l \\ l \end{vmatrix}} \cdot \begin{pmatrix} l \\ l \end{pmatrix} 3e_0^2 - 1 \begin{vmatrix} l \\ l \end{vmatrix} = \frac{3\lambda^2 - l(l+1)}{(2l-1)\,l} \cdot \frac{-2l}{2l+3} \ .$$

Sodann gilt

$$\overset{\lambda = l}{\underset{l-q+1}{\sum}} [3\lambda^2 - l(l+1)] = \frac{q}{4}\,(2l+1-q)\,(2l+1-2q) \ , \tag{A3.6}$$

---
[*]
$$\begin{pmatrix} l \\ l \end{pmatrix} 3e_0^2 - 1 \begin{vmatrix} l \\ l \end{vmatrix} = \oint d\boldsymbol{e}^2\ 2\sqrt{4\pi/5}\ Y_{20}(\boldsymbol{e}) \cdot |Y_{ll}(\boldsymbol{e})|^2$$
$$= (2J_l - 3J_{l+1}) : J_l = -2l/(2l+3) \ .$$

Dabei wurde folgende Abkürzung und Gleichung benutzt:

$$J_l \equiv \int\limits_0^\pi d\vartheta\,(\sin\vartheta)^{2l+1} = \frac{2l}{2l+1} \cdot \frac{2l-2}{2l-1} \cdot \ldots \cdot \frac{4}{5} \cdot \frac{2}{3} \cdot 2 \ .$$

wovon man sich am einfachsten durch vollständige Induktion überzeugt. Folglich ist

$$\sum_{l-q+1}^{\lambda=l} \left(\begin{array}{c}l\\\lambda\end{array}\middle| 3e_0^2 - 1\middle|\begin{array}{c}l\\\lambda\end{array}\right) = \frac{-2}{(2l-1)(2l+3)} \cdot \frac{q}{4} \times$$
$$\times (2l+1-q)(2l+1-2q) \tag{A3.7}$$

und also

$$\xi = \frac{1}{2} \cdot \frac{-q(2l+1-q)(2l+1-2q)}{2(2l-1)(2l+3)} \cdot \frac{1}{L(2L-1)S}, \tag{A3.8}$$

was nach (3.1) mit (6.5) übereinstimmt.

II. Löcherkonfigurationen. Diesen Fall kann man wieder auf den vorhergehenden zurückführen: Man nehme die Ersetzungen

$$\sum_{k=1}^{q} \to - \sum_{\bar{k}=1}^{\bar{q}}, \quad l_k \to -l_{\bar{k}}, \quad s_k = -s_{\bar{k}} \tag{A3.9}$$

vor (mit $L \to L$ und $S \to S$). Damit geht $\xi(q)$ in $\xi(\bar{q})$, d. h. $\xi(L, S)$ geht in sich über. Damit ist (6.5) auch in diesem Falle bewiesen.

Literatur

ABRAGAM, A.: The Principles of Nuclear Magnetism. Oxford 1961.
—, and M. H. L. PRYCE: Proc. Roy. Soc. London A **205**, 135 (1951)
BODENSTEDT, E.: Fortschr. Phys. **10**, 321 (1962)
BREIT, G.: Nature **122**, 649 (1928)
— Phys. Rev. **34**, 553 (1929)
EDMONDS, A. R.: Angular Momentum in Quantum Mechanics. Princeton 1957
FANO, U., and G. RACAH: Irreducible Tensorial Sets, New York 1959
FERMI, E.: Z. Physik **60**, 320 (1930)
FOCK, V.: Z. Physik **61**, 126 (1930)
FOLDY, L. L., and S. A. WOUYTHUYSEN: Phys. Rev. **78**, 29 (1950)
GOEPPERT-MAYER, M.: Phys. Rev. **60**, 184 (1941)
GOLDRING, G., and P. R. SCHARENBERG: Phys. Rev. **110**, 701 (1958)
HARTREE, D. R.: The Calculation of Atomic Structure. Princeton 1957
HELMERS, K.: Z. Physik **154**, 310 (1959)
HUND, F.: Linienspektren und periodisches System der Elemente (1927)
JUDD, B. R., and I. LINDGREN: Phys. Rev. **122**, 1802 (1961)
KANAMORI, I., and K. SUGIMOTO: J. Phys. Soc. Japan **13**, 754 (1958)
KARLSSON, E., E. MATTHIAS, and K. SIEGBAHN: Perturbed Angular Correlations (1964)
KOPFERMANN, H.: Kernmomente, 2. Aufl. (1956)
MARGENAU, H.: Phys. Rev. **57**, 383 (1940)
VAN VLECK, J. H.: Electric and Magnetic Susceptibilities (1932)

Abgeschlossen am 11. Juni 1964

Dipl. Phys. W. DONNER und Prof. Dr. G. SÜSSMANN
Institut für Theoretische Physik der Universität
Frankfurt/Main, Robert-Mayer-Straße 6—8

Group Theory and Spectroscopy *

Giulio Racah

The Hebrew University
Jerusalem

Contents

  * These notes are based on a series of seminar lectures given during the 1951 Spring term at the Institute for Advanced Study, Princeton. Owing to limitations of time only particular topics were considered, and there is no claim to completeness. The notes are by EUGEN MERZBACHER and DAVID PANK.

  The lectures were reproduced by CERN (61—68) in March 1961. We are very grateful to CERN and to Prof. RACAH for the permission of reprinting the lectures.

These lectures will treat the applications of group theory to problems of spectroscopy and nuclear structure. While developing the mathematical tools for this purpose, we shall occasionally forego the elaboration of a rigorous proof. In such cases, references will be quoted.

Part I. General notions on continuous groups

1. Continuous groups and infinitesimal groups

We start with a set of n variables x_0^i $(i = 1, \ldots, n)$, which may be regarded as coordinates of a point in a certain space. Consider now the set of equations

$$x^i = f^i(x_0^1, \ldots, x_0^n; a^1, \ldots, a^r) \qquad (i = 1, \ldots, n) , \quad (1)$$

in which the a^ϱ appear as a set of r independent parameters. Omitting indices, we shall write this and similar relations in the form

$$x = f(x_0; a) \quad \text{or} \quad x = S_a x_0 . \tag{1'}$$

These equations define a set S of transformations, depending on the parameters a, which map the point x_0 onto x. We shall assume that the f^i have all the required derivatives, and that the f^i depend *essentially* on the parameters, i. e. that no two transformations with different parameters are the same for all values of x_0, so that r is the smallest number of parameters needed to specify the transformations completely and uniquely.

The set of transformations f is said to form a group if it obeys the following two conditions:

i) The result of performing successively any two transformations of the set is another transformation belonging to this set. Formally, if $x = f(x_0; a)$ and $x' = f(x; b)$ then there exists a set of parameters c^ϱ

$$c^\varrho = \varphi^\varrho(a; b) \tag{2}$$

such that

$$x' = f(x; b) = f(f(x_0, a); b) = f(x_0; c) = f(x_0; \varphi(a; b)) . \tag{3}$$

ii) Corresponding to every transformation there exists a unique inverse, which also belongs to the set: Given equation (1) there exists a set of parameters $\bar{a}$ such that $x_0 = f(x, \bar{a})$.

The uniqueness of $\bar{a}$ is guaranteed if the Jacobian of the transformation does not vanish:

$$\left| \frac{\partial f}{\partial x_0} \right| \neq 0 . \tag{4}$$

Transforming x_0 onto x and then inversely x back to x_0, we obtain according to i) a transformation which belongs to the group and is characterized by the set of parameters a_0. Since the transformation depends on the parameters in an essential way, the a_0 so constructed cannot depend on the particular value of the parameters from which we started. The transformation $f(x, a_0)$ is called the identity.

Since it imposes no restriction, we shall take

$$a_0^\varrho = 0 \qquad\qquad (\varrho = 1, \ldots, r) .$$

r is called the *order* of the group. (Note that this usage is different from that found in the theory of finite groups.)

We also remind the reader of the following definitions:

A mapping of one group onto another is said to be *homomorphic* or a *homomorphism* if it preserves the operation of group multiplication. We call such a mapping an *isomorphism* if, in addition, the correspondence between elements of the two groups is one-to-one. Since the combination law of the transformations (1) is given in terms of the parameters a, there can be transformations corresponding to different values of n which are homomorphic or even isomorphic.

A group of linear transformations which is homomorphic with a given group is called a *representation* of this group.

The fundamental idea of Sophus Lie's theory of continuous groups is to consider not the whole of a group, but that part of it which lies near the identity, consisting of the so-called *infinitesimal transformations*. Thus, instead of the finite displacement of a point under a transformation, we consider the application of successive infinitesimal displacements — we think of a generalized velocity field describing the motion of a point from its original position x_0 to its final position x.

We have now two equivalent expressions for x

a) $\qquad\qquad x = f(x_0; a) \quad \text{or} \quad b) \; x = f(x; 0) .$ $\qquad\qquad$ (5)

Corresponding to these we can represent in either of two ways a transformation, as a result of which the new components of x differ infinitesimally from the old ones — by differentiation of (5a) or by introducing a parameter of infinitesimal size in (5b)

$$x + \mathrm{d}x = f(x_0; a + \mathrm{d}a) \quad \text{or} \quad x + \mathrm{d}x = f(x, \delta a)$$

or (employing the summation convention)

$$\mathrm{d}x = \frac{\partial f(x_0, a)}{\partial a^\sigma} \, \mathrm{d}a^\sigma \quad \text{or} \quad \mathrm{d}x = \left(\frac{\partial f(x, a)}{\partial a^\sigma} \right)_{a=0} \delta a^\sigma . \qquad (6)$$

The last may be written

$$\mathrm{d}x^i = u_\sigma^i(x) \, \delta a^\sigma , \quad u_\sigma^i(x) = \left(\frac{\partial f^i(x, a)}{\partial a^\sigma} \right)_{a=0} \qquad (7)$$

which defines the "velocity field" $u_\sigma^i(x)$ mentioned above. In the notation of (2) we may write

$$a + \mathrm{d}a = \varphi(a; \delta a) .$$

Since it follows from (2) and (5) that $\varphi(a; 0) = a$, we have

$$a + \mathrm{d}a = a + \left(\frac{\partial \varphi(a; b)}{\partial b^\tau}\right)_{b=0} \delta a^\tau .$$

Thus, $\mathrm{d}a$ is a linear combination of δa

$$\mathrm{d}a^\varrho = \mu_\tau^\varrho(a)\, \delta a^\tau , \quad \mu_\tau^\varrho(a) = \left(\frac{\partial \varphi^\varrho(a, b)}{\partial b^\tau}\right)_{b=0} . \tag{8}$$

Solving for δa, we get

$$\delta a^\sigma = \lambda_\varrho^\sigma(a)\, \mathrm{d}a^\varrho \tag{8'}$$

where

$$\lambda \mu = 1 , \quad \text{i. e.} \quad \lambda_\varrho^\sigma \mu_\tau^\varrho = \delta_\tau^\sigma . \tag{9}$$

From (6), (7) and (8') we get the first fundamental formula

$$\frac{\partial x^i}{\partial a^\varrho} = u_\tau^i(x)\, \lambda_\varrho^\tau(a) . \tag{A}$$

If u is to represent the velocity field of a transformation (1), equation (A) must be completely integrable, i. e. it must be capable of admitting solutions with n arbitrary constants x_0. The integrability condition $\frac{\partial^2 x^i}{\partial a^\sigma \partial a^\varrho} = \frac{\partial^2 x^i}{\partial a^\varrho \partial a^\sigma}$ becomes

$$\left(u_\varkappa^j \frac{\partial u_\nu^i}{\partial x^j} - u_\nu^j \frac{\partial u_\varkappa^i}{\partial x^j}\right) \lambda_\varrho^\varkappa \lambda_\sigma^\nu + u_\tau^i \left(\frac{\partial \lambda_\sigma^\nu}{\partial a^\varrho} - \frac{\partial \lambda_\varrho^\nu}{\partial a^\sigma}\right) = 0 ,$$

and, using (9), this gives

$$u_\varkappa^j \frac{\partial u_\nu^i}{\partial x^j} - u_\nu^j \frac{\partial u_\varkappa^i}{\partial x^j} = c_{\varkappa\nu}^\tau(a)\, u_\tau^i \tag{10}$$

where

$$c_{\varkappa\nu}^\tau(a) = \left(\frac{\partial \lambda_\varrho^\tau}{\partial a^\sigma} - \frac{\partial \lambda_\sigma^\tau}{\partial a^\varrho}\right) \mu_\varkappa^\varrho \mu_\nu^\sigma . \tag{11}$$

Since u is independent of a, differentiation of (10) by a^ϱ gives

$$\frac{\partial c_{\varkappa\nu}^\tau(a)}{\partial a^\varrho}\, u_\tau^i = 0 .$$

But the a's have been assumed essential, so that by (7) the u's are linearly independent, hence the c's are independent of a. Equation (10) is

$$u_\varkappa^j \frac{\partial u_\nu^i}{\partial x^j} - u_\nu^j \frac{\partial u_\varkappa^i}{\partial x^j} = c_{\varkappa\nu}^\tau\, u_\tau^i \tag{B_1}$$

and from (11)

$$\frac{\partial \lambda_\varrho^\tau}{\partial a^\sigma} - \frac{\partial \lambda_\sigma^\tau}{\partial a^\varrho} = c_{\varkappa\nu}^\tau\, \lambda_\varrho^\varkappa \lambda_\sigma^\nu . \tag{B_2}$$

(B_1) is a necessary condition on the velocity field if the latter is to generate a group, and (B_2) is a corresponding restriction on the manner in which the a's combine.

An infinitesimal transformation on the x induces on any function $F(x)$ a variation

$$\mathrm{d}F(x) = \frac{\partial F}{\partial x^i}\, \mathrm{d}x^i = \delta a^\sigma u_\sigma^i \frac{\partial F}{\partial x^i} \equiv \delta a^\sigma X_\sigma F \tag{12}$$

where

$$X_\sigma = u_\sigma^i(x) \frac{\partial}{\partial x^i} . \tag{13}$$

(12) shows that every infinitesimal transformation of $F(x)$ is generated by a linear combination of the operators X which are called the infinitesimal operators of the group S. From (B_1) it follows that they satisfy the relation

$$X_\varrho X_\sigma - X_\sigma X_\varrho \equiv [X_\varrho X_\sigma] = c_{\varrho\sigma}^\tau X_\tau \, . \tag{14}$$

Evidently

$$c_{\varrho\sigma}^\tau = -c_{\sigma\varrho}^\tau \, . \tag{C_1}$$

Substituting (14) into the Jacobi identity

$$[[X_\varrho X_\sigma]\, X_\tau] + [[X_\sigma X_\tau]\, X_\varrho] + [[X_\tau X_\varrho]\, X_\sigma] = 0$$

we get

$$c_{\varrho\sigma}^\mu\, c_{\mu\tau}^\nu + c_{\sigma\tau}^\mu\, c_{\mu\varrho}^\nu + c_{\tau\varrho}^\mu\, c_{\mu\sigma}^\nu = 0 \, . \tag{C_2}$$

We have shown that equations (C) are implied if the f^i form a group. That the converse to this statement holds is the content of the three fundamental theorems of Lie, which we shall not prove. They state that

I. If there exist $f^i = x^i$ satisfying (A), then they form a group.

II. If there exist u's satisfying (B_1), then there exist λ's, determined within isomorphism, which satisfy (B_2), so that equation (A) is integrable.

III. For every set of c's satisfying (C), there exist u's satisfying (B_1).

We shall write an infinitesimal transformation of the group S in the form $S_a = 1 + \delta a^\sigma X_\sigma$, where δa^σ is an infinitesimal quantity defined to be of the first order. If we combine two such transformations, we get

$$S_a S_b = (1 + \delta a^\varrho X_\varrho)\, (1 + \delta b^\sigma X_\sigma) = 1 + \delta a^\varrho X_\varrho + \delta b^\sigma X_\sigma \, ,$$

where the first non-vanishing infinitesimal terms have been retained. Thus, to the operation of multiplication in S corresponds addition in the infinitesimal group of S. If the first order quantities vanish, we have to consider quantities of higher order. But the second theorem of Lie implies that in this connection we need never go beyond the second order of infinitesimals — i. e. we have only to worry about *commutators*, which are expressions of the form $S_a S_b S_a^{-1} S_b^{-1}$, and to ask that the corresponding infinitesimal operator of the second order, $\delta a^\varrho\, \delta b^\sigma [X_\varrho X_\sigma]$ be contained in the linear manifold of infinitesimal operators.

2. Parameter groups and adjoint group

On comparing (8) with (7), we see that a close formal analogy exists between the functions μ and u. In fact, the $\varphi(a; b)$ of (2) which connect the parameters according to the composition law of the group may themselves be considered as defining a group in the same way as does (1')

$$a'^\varrho = \varphi^\varrho(a; b) \, .$$

This relation can be regarded as a mapping of the a onto the a' according to a transformation whose parameter is b. We shall prove that these transformations form a group P_1, which is isomorphic with S and is called the *first parameter group*. Indeed, if $a' = \varphi(a; b)$ and $a'' = \varphi(a'; b)$ then

$$a'' = \varphi(\varphi(a; b); c) = \varphi(a; \varphi(b; c))$$

where the last equality follows from the associative property of the transformation f^i. We thus see that the law of compositions is the same for the first parameter group and the original group of transformations f^i.

The analogous group of transformations on the argument b of $\varphi(a; b)$ is called the *second parameter group* P_2. P_2 is anti-isomorphic with S, that is, it is isomorphic when the factors are taken in the reverse order. But since $(x\,y)^{-1} = y^{-1}\,x^{-1}$ and since a group contains x^{-1} if it contains x, the two are in fact isomorphic. Let a (or c) be a transformation belonging to the first (or second) parameter group. Let the operation of P_1 transform b into $b' = \varphi(a, b)$ and let the operation of P_2 transform b' into $b'' = \varphi(c, b')$. Then it is clear that

$$b'' = \varphi(c, b') = \varphi(c, \varphi(b, a)) = \varphi(\varphi(c, b), a) ,$$

hence every element of P_1 commutes with every element of P_2.

The μ's are the velocity field of P_1 [see equation (8)], and they define the infinitesimal operator

$$A_\tau = \mu_\tau^\varrho(a)\, \frac{\partial}{\partial a^\varrho} . \tag{15}$$

Correspondingly, in P_2

$$B_\tau = \bar{\mu}_\tau^\varrho(b)\, \frac{\partial}{\partial b^\varrho} . \tag{16}$$

Another operation which it is useful to consider is conjugation: Given an element S_a of S, to every element S_b of the group there corresponds an element $S_{b'} = S_a S_b S_a^{-1}$. The operation $b \to b'$ is a faithful mapping of the group onto itself which depends on S_a and which is called conjugation of S by S_a. Consider now the set of conjugations obtained by letting S_a run through all the elements of S. These conjugations themselves constitute a group of transformations, homomorphic to S, but not in general isomorphic with S. It is easily seen that isomorphism between S and the group of conjugations holds if and only if the identity is the only element of S which commutes with all elements of S.

If we regard the relation $x' = S_a x$ as a coordinate transformation, it is well known that the effect of operating with S_b on a function of x' is given in terms of x' by the operation of the conjugate of S_b by S_a acting on the same function of x

$$S_{b'}\, x' = S_a S_b\, x = (S_b\, x)' .$$

The conjugation gives the change in the parameters of an operation if this operation is considered in the new system of coordinates x'. The advantage of the group of conjugations over the parameter group is that the conjugate element $S_a S_\varepsilon S_a^{-1}$ is infinitesimal if S_ε is infinitesimal, irrespective of the magnitude of S_a.

If in the first system of coordinates S_ε is expressed by $S_\varepsilon = 1 + \varepsilon\, e^\varrho X_\varrho$, then, after the transformation with S_a the same transformation S_ε will be expressed by $1 + \varepsilon\, e'^\varrho X'_\varrho$, and hence

$$e'^\varrho X'_\varrho = e^\varrho X_\varrho \,. \tag{17}$$

The group of transformations $e^\varrho \to e'^\varrho$ is called the *adjoint group*. We wish to determine its infinitesimal operators, produced by transformations S_a in the neighborhood of the identity. With $S_a = 1 + \delta a^\sigma X_\sigma$ and $S_\varrho = 1 + \varepsilon\, X_\varrho$ we have

$$S'_\varrho = 1 + \varepsilon\, X'_\varrho = S_a S_\varrho S_a^{-1} = (S_a S_\varrho S_a^{-1} S_\varrho^{-1})\, S_\varrho = (1 + [\delta a^\sigma X_\sigma, \varepsilon\, X_\varrho]) \times$$
$$\times (1 + \varepsilon\, X_\varrho)$$

or

$$X'_\varrho - X_\varrho \equiv \mathrm{d}X_\varrho = \delta a^\sigma [X_\sigma, X_\varrho] = c^\tau_{\sigma\varrho}\, \delta a^\sigma X_\tau \,,$$

by (14). From (17), we have

$$\mathrm{d}e^\tau X_\tau = -e^\varrho\, \mathrm{d}X_\varrho = e^\varrho\, c^\tau_{\varrho\sigma}\, \delta a^\sigma X_\tau$$

or

$$\mathrm{d}e^\tau = e^\varrho\, c^\tau_{\varrho\sigma}\, \delta a^\sigma \,. \tag{18}$$

If E_σ are the infinitesimal operators of the adjoint group, we find by comparison of (18) with (7) and (13) that

$$E_\sigma = e^\varrho\, c^\tau_{\varrho\sigma}\, \frac{\partial}{\partial e^\tau} \,. \tag{19}$$

3. Subgroups, simple and semi-simple groups

a) A group is *Abelian* if all its elements commute. It follows from the correspondence between commutators and square brackets that for an Abelian group all square brackets, and consequently all structure constants, vanish

$$c^\tau_{\varrho\sigma} = 0 \,. \tag{20}$$

b) A *subgroup* of a group S is a subset of elements of S which satisfies the group postulates. Thus, if $X_1, X_2, \ldots, X_p$ are the infinitesimal operators of a subgroup, the structure constants of the group must satisfy the relations

$$c^\tau_{\varrho\sigma} = 0 \quad (\varrho, \sigma \leq p, \tau > p) \,. \tag{21}$$

c) An *invariant subgroup*, H, of a group S is a subgroup of S which contains all the conjugates (images) of its elements. Thus, with S_n, it contains $S_x S_n S_x^{-1}$ for any S_x in S. If so, it also contains the commutator $S_x S_n S_x^{-1} S_n^{-1}$. Thus, the square bracket connecting an infinitesimal element of H with any infinitesimal element of S must belong to H. If $X_1, X_2, \ldots, X_p$ are the infinitesimal operators of an invariant subgroup of S, the structure constants of S must satisfy

$$c^\tau_{\varrho\sigma} = 0 \quad (\varrho \leq p, \tau > p) \,. \tag{22}$$

d) A group is *simple* if it has no invariant subgroup besides the unit element.

e) A group is *semi-simple* if it has no Abelian invariant subgroups besides the unit element.

The distinction between groups which have Abelian invariant subgroups and those which do not have such subgroups is important, because Abelian subgroups, though apparently easiest to deal with, can actually be most troublesome from the point of view of representations, as the following example will show:

We consider the group of rectilinear motions in one dimension, in which the transformation $x' = x + a$ followed by $x'' = x' + b$ is equivalent to $x'' = x + a + b$. This group can be represented by square matrices of the second rank, in terms of which the composition law just given would read

$$\begin{pmatrix} 1 & a \\ 0 & 1 \end{pmatrix} \begin{pmatrix} 1 & b \\ 0 & 1 \end{pmatrix} = \begin{pmatrix} 1 & a+b \\ 0 & 1 \end{pmatrix}.$$

However, none of the matrices of this particular representation can be brought to diagonal form by a similarity transformation. This peculiar behavior is closely related to the Abelian property. Indeed, as we shall show later, semi-simple groups never exhibit it. Moreover, the physical applications in which we shall be interested will require the use only of semi-simple groups. We shall therefore from this point on restrict ourselves to the study of semi-simple groups. To this end, we must have a criterion for their identification.

Such a criterion can be formulated very simply in terms of a symmetrical tensor of the second rank which we construct from the $c^{\tau}_{\varrho\sigma}$

$$g_{\varrho\sigma} = c^{\mu}_{\varrho\lambda} c^{\lambda}_{\sigma\mu}. \tag{23}$$

If the group is semi-simple, then necessarily,

$$\det |g_{\varrho\tau}| \neq 0. \tag{24}$$

For suppose it possesses an Abelian invariant subgroup, the indices of whose elements are denoted by $\bar{\varrho}, \bar{\sigma}, \ldots$. Then,

$$
\begin{aligned}
g_{\varrho\bar{\sigma}} &= c^{\mu}_{\varrho\lambda} c^{\lambda}_{\bar{\sigma}\mu} \\
&= c^{\mu}_{\varrho\lambda} c^{\bar{\lambda}}_{\bar{\sigma}\mu} \qquad \text{by (22)} \\
&= c^{\bar{\mu}}_{\varrho\bar{\lambda}} c^{\bar{\lambda}}_{\bar{\sigma}\bar{\mu}} \qquad \text{by (22)} \\
&= 0 \qquad \text{by (20)} .
\end{aligned}
$$

That the condition (24) is sufficient as well as necessary has been shown by CARTAN.

We can use the tensor $g_{\mu\nu}$ to define a relation of orthogonality between contravariant vectors or to form new tensors by lowering of indices. As an example,

$$c_{\varrho\sigma\lambda} = c^{\tau}_{\varrho\sigma} g_{\tau\lambda} \tag{25}$$

and this new tensor is totally antisymmetric, for by (23),

$$
\begin{aligned}
c_{\varrho\sigma\lambda} &= c^{\tau}_{\varrho\sigma} c^{\nu}_{\tau\mu} c^{\mu}_{\lambda\nu} \\
&= -c^{\tau}_{\sigma\mu} c^{\nu}_{\tau\varrho} c^{\mu}_{\lambda\nu} - c^{\tau}_{\mu\varrho} c^{\nu}_{\tau\sigma} c^{\mu}_{\lambda\nu} \qquad \text{by } (C_2) \\
&= c^{\tau}_{\sigma\mu} c^{\nu}_{\varrho\tau} c^{\mu}_{\lambda\nu} + c^{\tau}_{\mu\varrho} c^{\nu}_{\tau\sigma} c^{\mu}_{\nu\lambda} \qquad \text{by } (C_1) .
\end{aligned}
$$

3*

The last line has the desired property, since it is invariant under cyclic permutation of the indices and is, by construction, skew in ϱ and σ.

If the group S is semi-simple, then Cartan's criterion (24) implies that we can form from $g_{\varrho\sigma}$ the reciprocal tensor $g^{\varrho\sigma}$ which can be used to raise indices and define orthogonality between covariant vectors.

As an example of the foregoing we consider the group of rigid motions in three dimensions, consisting of rotations and translations. The infinitesimal rotations are generated by operators L_j $(j = 1, 2, 3)$ satisfying

$$[L_1 L_2] = i L_3 \text{ etc.} \tag{26}$$

and the infinitesimal displacements by operators $P_1 = L_4$, $P_2 = L_5$, $P_3 = L_6$ which commute among themselves but which satisfy

$$[L_1 L_5] = i L_6 \text{ etc.}$$

so that the only non-vanishing structure constants are

$$c_{12}^3 = c_{23}^1 = c_{15}^6 = c_{26}^4 = c_{34}^5 = c_{31}^2 = c_{61}^5 = c_{42}^6 = c_{53}^4 = i$$

plus a corresponding list given by (C_1). For the $g_{\varrho\sigma}$ we find

$$g_{11} = g_{22} = g_{33} = 4, \quad g_{44} = g_{55} = g_{66} = 0, \quad g_{\varrho\sigma} = 0 \ (\varrho \neq \sigma) \ .$$

The determinant det $g_{\varrho\sigma}$ vanishes, as required by Cartan's criterion, since the translations form an Abelian invariant subgroup.

If we consider only the group of three-dimensional rotations, defined by (26), we find that it is simple and that the metric tensor is

$$g_{\varrho\sigma} = 2 \delta_{\varrho\sigma} \ . \tag{27}$$

Part II. Classification of the semi-simple groups

1. The standard form of the infinitesimal group

In order to obtain a standard coordinate system for the set of infinitesimal operators of a semi-simple group we consider an eigenvalue problem of the form

$$[A \ X] = \varrho X \tag{28}$$

where A is a fixed arbitrary infinitesimal operator $A = a^\mu X_\mu$ while $X = x^\nu X_\nu$ is an *eigenvector* corresponding to the eigenvalue ϱ. Using (14) we can write (28) explicitly as

$$a^\mu x^\nu c_{\mu\nu}^\tau X_\tau = \varrho \, x^\tau X_\tau \ .$$

Since the infinitesimal operators are linearly independent it follows that

$$(a^\mu c_{\mu\nu}^\tau - \varrho \, \delta_\nu^\tau) \, x^\nu = 0 \ . \tag{29}$$

From (29) we get the secular equation

$$\det (a^\mu c_{\mu\nu}^\tau - \varrho \, \delta_\nu^\tau) = 0 \ . \tag{30}$$

If there exist r linearly independent eigenvectors, they can be used as a basis for a coordinate system in the r-dimensional space. However, generally, r linearly independent eigenvectors may not exist if the secular equation has degenerate roots. Usually, in physical problems, conditions like hermiticity or symmetry of the matrix insure the existence of r linearly independent eigenvectors. But for semi-simple infinitesimal groups CARTAN has shown that if A is chosen so that the secular equation (30) has the maximum number of different roots, then only $\varrho = 0$ is degenerate; and that if l be the multiplicity of this root, there are corresponding to this root l linearly independent eigenvectors $H_1, \ldots, H_l$ which commute with each other. l is called the *rank* of the semi-simple group. (Since A commutes with itself, the rank of a semi-simple group is at least one.)

We shall use Latin indices $1, \ldots, l$ for the coordinates in the subspace of dimension l, spanned by the H_i, while Greek indices $\alpha, \ldots, \nu$ will be employed for the $r - l$ dimensional subspace which is spanned by the eigenvectors $E_\alpha, \ldots, E_\nu$ corresponding to the non-vanishing distinct roots $\alpha, \ldots, \nu$. For the latter indices the summation convention will be suspended. The three indices ϱ, σ, τ will be used to refer to the whole r-dimensional space.

The basic vectors H_i and E_α are defined by the relations

$$[A \, H_i] = 0 \qquad\qquad (i = 1, \ldots, l) , \qquad (31)$$

$$[A \, E_\alpha] = \alpha E_\alpha . \qquad\qquad (32)$$

Further, since A is an eigenvector of (28) with eigenvalue zero, it can be written in the form

$$A = \lambda^i \, H_i . \qquad (33)$$

We shall now discuss the commutators of H's and E's, in order to obtain information about the $c^\tau_{\varrho\sigma}$. First, from Cartan's theorem, we have

$$[H_i \, H_k] = 0 \quad \text{or} \quad c^\tau_{ik} = 0 . \qquad (34)$$

Second, we consider $[H_i \, E_\alpha]$. To do this we write

$$[A \, [H_i \, E_\alpha]] + [H_i \, [E_\alpha A]] + [E_\alpha \, [A \, H_i]] = 0 .$$

By (31) and (32) this is

$$[A \, [H_i \, E_\alpha]] = \alpha [H_i \, E_\alpha] . \qquad (35)$$

Thus $[H_i \, E_\alpha]$ is an eigenvector of (28) belonging to $\varrho = \alpha$, and since these eigenvectors are not degenerate, we must have

$$[H_i \, E_\alpha] = \alpha_i \, E_\alpha \quad \text{or} \quad c^\tau_{i\alpha} = \alpha_i \, \delta^\tau_\alpha . \qquad (36)$$

From (32), (33) and (36) follows that

$$\alpha = \lambda^i \, \alpha_i . \qquad (37)$$

From here on the letter α or the term "root" will be used to denote either the form (37) or the vector with covariant components α_i in the l-dimensional space.

Finally, to find $[E_\alpha \, E_\beta]$, we form

$$[A \, [E_\alpha \, E_\beta]] + [E_\alpha \, [E_\beta A]] + [E_\beta \, [A \, E_\alpha]] = 0 .$$

By (32), this is
$$[A [E_\alpha E_\beta]] = (\alpha + \beta) [E_\alpha E_\beta] . \tag{38}$$
Hence $[E_\alpha E_\beta]$ belongs as eigenvector to the root $\alpha + \beta$, if $\alpha + \beta$ is a root, and vanishes, if $\alpha + \beta$ is not a root. If $\alpha + \beta$ is a non-vanishing root, we shall write
$$[E_\alpha E_\beta] = N_{\alpha\beta} E_{\alpha+\beta} \quad \text{or} \quad c_{\alpha\beta}^{\alpha+\beta} = N_{\alpha\beta} . \tag{39}$$
If $\beta = -\alpha$, then evidently we have
$$[E_\alpha E_{-\alpha}] = c_{\alpha-\alpha}^i H_i , \tag{40}$$
and for the rest,
$$c_{\alpha\beta}^\tau = 0 . \qquad (\tau \neq \alpha + \beta) \tag{41}$$

We shall now show that if α is a root, then $-\alpha$ is also a root. This is done by forming the tensor $g_{\alpha\tau}$. The restrictions (36), (40) and (41), when applied to (23), give
$$g_{\alpha\tau} = c_{\alpha i}^\alpha c_{\tau\alpha}^i + \sum_{\beta \neq -\alpha} c_{\alpha\beta}^{\alpha+\beta} c_{\tau\alpha+\beta}^\beta + c_{\alpha-\alpha}^i c_{\tau i}^{-\alpha} . \tag{42}$$
But by (36) and (41), each term on the right of (42) exists only when $\tau = -\alpha$, so that
$$g_{\alpha\tau} = 0 \qquad (\tau \neq -\alpha). \tag{43}$$
Thus, if $-\alpha$ is not a root, Cartan's criterion (24) for the semi-simple groups is violated. By a suitable normalization of E_α we may set
$$g_{\alpha-\alpha} = 1 , \tag{44}$$
and we can order our basis so that the tensor $g_{\varrho\sigma}$ is written in the form
$$g_{\varrho\sigma} = \begin{pmatrix} g_{ik} & & 0 \\ \hline & \begin{matrix} 0 & 1 \\ 1 & 0 \end{matrix} & \\ 0 & & \begin{matrix} 0 & 1 \\ 1 & 0 \end{matrix} \\ & 0 & \ddots \end{pmatrix} . \tag{45}$$
Since $\det g_{\varrho\sigma}$ is the product of the elementary determinants, it follows from (24) that
$$\det g_{ik} \neq 0 . \tag{46}$$
Further,
$$g_{ik} = \sum_\alpha c_{i\alpha}^\alpha c_{k\alpha}^\alpha = \sum_\alpha \alpha_i \alpha_k . \tag{47}$$
It may be noted that the g_{ik} defined by (47) has a non-vanishing determinant only if the vectors α span the entire l-dimensional space. g_{ik} will be used as the metric tensor for this space.

Using the inverse tensor we can now establish the following useful identity
$$\begin{aligned} c_{\alpha-\alpha}^i &= g^{ik} c_{\alpha-\alpha k} \\ &= g^{ik} c_{k\alpha-\alpha} \quad \text{by antisymmetry in subscripts,} \\ &= g^{ik} c_{k\alpha}^\alpha \quad \text{by (44)} \\ &= g^{ik} \alpha_k \equiv \alpha^i \quad \text{by (36),} \end{aligned} \tag{48}$$

so that (40) can be written as

$$[E_\alpha E_{-\alpha}] = \alpha^i H_i ,\qquad (49)$$

where the α^i are the contravariant components of the vector α. Collecting (34), (36), (39) and (49), we have for the standard forms of the commutation relations

$$[H_i H_k] = 0$$
$$[H_i E_\alpha] = \alpha_i E_\alpha$$
$$[E_\alpha E_\beta] = N_{\alpha\beta} E_{\alpha+\beta} \quad \text{when } \alpha + \beta \text{ is a non-vanishing root}$$
$$[E_\alpha E_{-\alpha}] = \alpha^i H_i .$$

$$(50)$$

As an example of the foregoing, we can take the operators of rotation in three dimensions, generated by L_1, L_2, L_3 such that

$$[L_1 L_2] = i L_3, \text{ etc.} \qquad (51\,\text{a})$$

If we take A equal to L_3, the two relations

$$[L_3, L_1 \pm i L_2] = \pm (L_1 \pm i L_2) \qquad (51\,\text{b})$$

show that $L_1 \pm i L_2$ are eigenvectors corresponding to $\varrho = \pm 1$. Use of the normalization condition (44) yields

$$H_1 = L_3 , \quad E_1 = \frac{L_1 + i L_2}{\sqrt{2}} , \quad E_{-1} = \frac{L_1 - i L_2}{\sqrt{2}} . \qquad (51\,\text{c})$$

2. Properties of the roots

We shall now prove the following

Theorem: If α and β are roots, then $\dfrac{2(\alpha\beta)}{(\alpha\alpha)}$ is an integer and $\beta - \dfrac{2(\alpha\beta)}{(\alpha\alpha)} \alpha$ is also a root*.

This theorem is to hold for arbitrary α and β, but we shall start by restricting β to be some root, γ, such that $\alpha + \gamma$ is not a root. According to (50) we can generate a set of operators

$$[E_{-\alpha} E_\gamma] = N_{-\alpha\gamma} E_{\gamma-\alpha} \equiv E'_{\gamma-\alpha}$$
$$[E_{-\alpha} E'_{\gamma-\alpha}] = E'_{\gamma-2\alpha}$$
$$\cdot \cdot \cdot \cdot \cdot \cdot \cdot \cdot \cdot \cdot \cdot \cdot$$
$$[E_{-\alpha} E'_{\gamma-j\alpha}] = E'_{\gamma-(j+1)\alpha}$$

$$(52)$$

where the primes indicate that, for the moment, we are not interested in the normalization of the E_β. Since there is only a finite number of E_β, this process must eventually stop after, say, g steps. Thus

$$[E_{-\alpha} E'_{\gamma-g\alpha}] = E'_{\gamma-(g+1)\alpha} = 0 . \qquad (53)$$

According to (39), $E_{\gamma-j\alpha}$ may be obtained again by an equation of the form

$$[E_\alpha E'_{\gamma-(j+1)\alpha}] = \mu_{j+1} E'_{\gamma-j\alpha} . \qquad (54)$$

---

* We use the notation $(\alpha\beta)$ for the scalar product $\alpha_i \beta^i$.

In order to evaluate the coefficients μ_{j+1} we eliminate $E'_{\gamma-(j+1)\alpha}$ from (52) and (54), thus finding

$$\mu_{j+1} E'_{\gamma-j\alpha} = - [E'_{\gamma-j\alpha} [E_\alpha E_{-\alpha}]] - [E_{-\alpha} [E'_{\gamma-j\alpha} E_\alpha]]$$
$$\text{by Jacobi's identity},$$

$$= - [E'_{\gamma-j\alpha}, \alpha^k H_k] + \mu_j [E_{-\alpha} E'_{\gamma-(j-1)\alpha}],$$
$$\text{by (50) and (54)}.$$

The use of (50) and (52) gives at once a recurrence relation for the μ_j

$$\mu_{j+1} = \mu_j + (\alpha\gamma) - j(\alpha\alpha). \tag{55}$$

This relation holds only for $j \geq 1$, as μ_0 is not defined by (54); however, the preceding argument shows that (55) can be extended to hold also for $j = 0$ if we define

$$\mu_0 = 0. \tag{56}$$

From (55) and (56) we obtain immediately

$$\mu_j = j(\alpha\gamma) - \frac{j(j-1)}{2}(\alpha\alpha). \tag{57}$$

It follows from (53) and (54) that $\mu_{g+1} = 0$, whence we have

$$(\alpha\gamma) = \frac{1}{2} g(\alpha\alpha), \tag{58}$$

where g is, by definition, a non-negative integer. Introducing (58) into (57) we get

$$\mu_j = \frac{j(g-j+1)}{2}(\alpha\alpha). \tag{59}$$

If $(\alpha\alpha)$ were zero for some root α, this root would, according to (58), be orthogonal to every root. But, as the roots span the entire l-dimensional space, this would contradict (46). Hence we can write

$$g = \frac{2(\alpha\gamma)}{(\alpha\alpha)}, \tag{60}$$

and we have proved that if α and γ are roots and $\alpha + \gamma$ is not a root, then there exists a *string* of roots,

$$\gamma, \gamma - \alpha, \ldots, \gamma - \frac{2(\alpha\gamma)}{(\alpha\alpha)}\alpha \equiv \gamma - g\alpha, \tag{61}$$

which is invariant under reflection with respect to the hyperplane through the origin perpendicular to the vector α. To return to the theorem which is to be proved, we note that for any root β, there exists some integer $j \geq 0$, such that $\beta + j\alpha$ is a root but $\beta + (j+1)\alpha$ is not. We can now set $\beta + j\alpha = \gamma$ in the above discussion, so that the string (61) can be written

$$\beta + j\alpha, \beta + (j-1)\alpha, \ldots, \beta, \ldots, \beta - k\alpha \qquad (j+k=g), \tag{62}$$

and as

$$2(\alpha\beta) = 2(\alpha\gamma) - 2j(\alpha\alpha) = (g-2j)(\alpha\alpha),$$

$\frac{2(\alpha\beta)}{(\alpha\alpha)}$ is an integer, and $\beta - \frac{2(\alpha\beta)}{(\alpha\alpha)}\alpha$ is contained in the string (62).

In (39) we introduced a set of coefficients $N_{\alpha\beta}$, but we have yet to see whether some of them may not vanish. This we can do with the aid of the theorem just proved. Assuming that with α and β, $\alpha + \beta$ is a root we evaluate

$$[E_{-\alpha} E_{\alpha+\beta}] = [E_{-\alpha} E_{\gamma-(j-1)\alpha}] = N_{-\alpha\alpha+\beta} E_{\gamma-j\alpha} .$$

With this we form

$$N_{-\alpha\alpha+\beta} [E_{\alpha} E_{\gamma-j\alpha}] = N_{\alpha\beta} N_{-\alpha\alpha+\beta} E_{\gamma-(j-1)\alpha} = \mu_j E_{\gamma-(j-1)\alpha} , \text{ by (54)} .$$

Equations (59) and (62) now tell us that

$$N_{\alpha\beta} N_{-\alpha\alpha+\beta} = \mu_j = \frac{j(k+1)}{2} (\alpha\,\alpha) , \tag{63}$$

and from this it is evident that $N_{\alpha\beta} \neq 0$ if $\alpha + \beta$ is a root and therefore $j \geq 1$.

It follows from this that if α is a root, 2α cannot be one, since E_{α} commutes with itself. From this, $k\alpha$ cannot be a root for any positive integer k, since if it were, it would determine a string which would contain 2α as an element. Hence, any string containing zero has only three elements α, 0, $-\alpha$.

We take now l linearly independent roots $\alpha^{(1)}, \ldots, \alpha^{(l)}$ as the basis of a new coordinate system in the l-dimensional space, and express all other root vectors as linear combinations

$$\beta = \sum_{k=1}^{l} b_k\, \alpha^{(k)} . \tag{64}$$

Multiplying (64) by $\alpha^{(i)}$ and dividing by $(\alpha^{(i)} \alpha^{(i)})$, which was shown to be different from zero, we get

$$\frac{(\beta\,\alpha^{(i)})}{(\alpha^{(i)} \alpha^{(i)})} = \sum_{k=1}^{l} b_k \frac{(\alpha^{(k)} \alpha^{(i)})}{(\alpha^{(i)} \alpha^{(i)})} .$$

Using the fundamental relation (60) we deduce readily that the new covariant components b_k must be real, rational, and even, by a change of scale, integral numbers.

This shows that for a suitable choice of the H_i the α_i are real, and this implies that g_{ik} is a positive definite matrix, since for any (real) x^i

$$g_{ik}\, x^i\, x^k = \sum_{\alpha} (\alpha\, x)^2 \geq 0 . \tag{65}$$

Hence the l-dimensional space has an ordinary Euclidean metric.

3. The vector diagrams

The graphical representation of the root vectors is called a vector diagram. SCHOUTEN derived restrictions on these diagrams from which all simple Lie groups can be found. The complete classification (already found algebraically by CARTAN) was obtained using this method by VAN DER WAERDEN, who showed also that to every vector diagram corresponds only one infinitesimal Lie group. Since the roots belong to a lattice which is invariant under a group of reflections, COXETER'S construction of all finite groups generated by reflection leads to a third method of classifying the simple groups. We shall here sketch the method of SCHOUTEN and VAN DER WAERDEN.

Suppose we have two roots, α and β, and let φ be the angle between them. We saw in the preceding section that

$$(\alpha\beta) = \frac{1}{2} m (\alpha\alpha) = \frac{1}{2} n (\beta\beta), \tag{66}$$

where m and n are integers. From this we get

$$\cos^2\varphi = \frac{(\alpha\beta)^2}{(\alpha\alpha)(\beta\beta)} = \frac{m n}{4}, \tag{67}$$

and from this we see that φ can have only the values $0°$, $30°$, $45°$, $60°$, and $90°$. From (66) we deduce that the ratios of the lengths of the two vectors are $\sqrt{3}$ for $30°$, $\sqrt{2}$ for $45°$, 1 for $60°$, and undetermined for $90°$. For $0°$ we know already that $\alpha = \beta$.

We want to construct every possible vector diagram which satisfies these conditions and those obtained in § 2. As CARTAN has shown that every semi-simple group is a direct product of simple groups, we shall be interested only in the diagrams of simple groups. We shall therefore not consider diagrams which can be split into mutually orthogonal parts, since evidently every such part corresponds to an invariant subgroup.

It is easy to see that the only possible two-dimensional diagrams are the ones drawn below. They are labelled by the letters which are traditional from Cartan's thesis; the numerical subscript denotes the rank of the group.

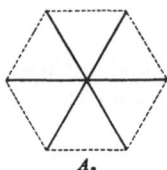   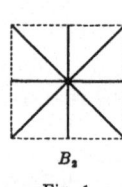   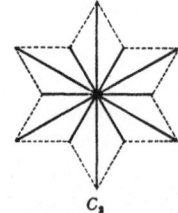

A_2 B_2 C_2

Fig. 1.

We shall now generalize these diagrams to l dimensions. In what follows, we shall denote by e_i a set of mutually orthogonal unit vectors.

A_l. The diagram A_2, above, may conveniently be regarded as consisting of all vectors of the form $e_i - e_k$ ($i, k = 1, 2, 3$). Generalizing to l dimensions, A_l is formed from $l + 1$ unit vectors e_i by forming the $l(l + 1)$ differences $e_i - e_k$. These will lie in the plane $\sum_{i=1}^{l+1} x^i = 0$. There are $l(l + 1)$ vectors, and adding to this the rank, which is the multiplicity of the root zero, we see that the group is of order $(l + 1)^2 - 1$.

B_l. We can generalize B_2 in l dimensions by constructing B_l out of all the vectors $\pm e_i$ and $\pm e_i \pm e_k$ ($i, k = 1, \ldots, l$). There are $2l^2$ vectors, and the order of the group is $l(2l + 1)$.

C_l. Another possible generalization of B_2, however, is to construct all vectors of the form $\pm 2e_i$ and $\pm e_i \pm e_k$ $(i, k = 1, \ldots, l)$. For $l = 2$, C_2 differs from B_2 only by rotation through $45°$. For $l > 2$, these diagrams are different from the B_l. C_l has the same order as B_l.

D_l. For $l > 2$, the diagram consisting of vectors $\pm e_i \pm e_k$ $(i, k = 1, \ldots, l)$ represents a simple group, which we shall call D_l. There are $2l(l-1)$ vectors, and the group is of order $l(2l-1)$. For $l = 2$, this construction gives only two orthogonal pairs of vectors and is therefore not simple. It may be noted that by a rotation of the axes given by

$$e_1' = \frac{1}{2}(e_1 + e_2 - e_3 - e_4)$$

$$e_2' = \frac{1}{2}(e_1 - e_2 + e_3 - e_4)$$

$$e_3' = \frac{1}{2}(e_1 - e_2 - e_3 + e_4) \tag{68}$$

$$e_4' = \frac{1}{2}(e_1 + e_2 + e_3 + e_4)$$

the vector diagram A_3 may be brought into coincidence with D_3.

VAN DER WAERDEN has shown that apart from these four classes of simple diagrams there are only five possible simple diagrams. One of them is G_2; the others are the following:

F_4. This diagram consists of the vectors of B_4 plus 16 more vectors $\frac{1}{2}(\pm e_1 \pm e_2 \pm e_3 \pm e_4)$. There are 48 vectors and the group is of order 52.

E_6 consists of the vectors of A_5, the vectors $\pm \sqrt{2}\, e_7$, and all the vectors

$$\frac{1}{2}(\pm e_1 \pm e_2 \pm e_3 \pm e_4 \pm e_5 \pm e_6) \pm \frac{e_7}{\sqrt{2}},$$

where in the first fraction we take three signs positive and three negative. There are 72 vectors and the group is of order 78.

E_7 consists of the vectors of A_7 and all the vectors

$$\frac{1}{2}(\pm e_1 \pm e_2 \pm e_3 \pm e_4 \pm e_5 \pm e_6 \pm e_7 \pm e_8),$$

where we take four signs positive and four negative. There are 126 vectors, and the group is of order 133.

E_8 consists of the vectors of D_8 and all the vectors

$$\frac{1}{2}(\pm e_1 \pm e_2 \pm e_3 \pm e_4 \pm e_5 \pm e_6 \pm e_7 \pm e_8),$$

with each sign occurring an even number of times. There are 240 vectors, and the group is of order 248.

The simplest realizations of the groups characterized by the vector diagrams A_l, B_l, C_l, D_l, are the classical groups, i. e. the special linear (unimodular), the orthogonal and the symplectic (complex) groups.

For the full linear group in $l + 1$ dimensions we may choose the infinitesimal operators

$$X_{ik} = x^i \frac{\partial}{\partial x^k}, \qquad (i, k = 1, \ldots, l+1) \quad (69)$$

with the commutation relations

$$[X_{ik} X_{mn}] = \delta_{km} X_{in} - \delta_{in} X_{mk} . \tag{70}$$

But the full linear group is not a semi-simple group; the operator $\sum_j X_{jj}$ commutes with every operator of the group, and the Abelian subgroup generated by this operator (i. e. the subgroup of the dilatations) is an invariant subgroup.

In order to have a semi-simple group we have to restrict ourselves to the unimodular subgroup (or "special" linear group) in $l + 1$ dimensions. Then the X_{ii} are no longer infinitesimal operators of the subgroup but should be replaced by

$$X'_{ii} = X_{ii} - \frac{1}{l+1} \sum_j X_{jj}, \tag{69'}$$

a change which does not affect the commutation relations (70). These operators correspond to the diagram A_l if we make the identification

$$X'_{ii} \equiv H_i, \quad X_{ik} \equiv E_{(e_i - e_k)} . \tag{71}$$

Although we have $l + 1$ operators H_i, only l of them are linearly independent, owing to the relation

$$\sum_{i=1}^{l+1} H_i = 0 . \tag{72}$$

For the orthogonal group in $2l + 1$ dimensions, which leaves the quadratic form

$$\sum_{k=-l}^{l} x^k x^{-k} = x_0^2 + 2 \sum_{k=1}^{l} x^k x^{-k}$$

invariant, we may choose the infinitesimal operators

$$X_{ik} = - X_{ki} = x^i \frac{\partial}{\partial x^{-k}} - x^k \frac{\partial}{\partial x^{-i}}, \quad (i, k = 0, \pm 1, \ldots, \pm l), \quad (73)$$

with the commutation relations

$$[X_{ik} X_{mn}] = \delta_{k+m} X_{in} - \delta_{k+n} X_{im} - \delta_{i+m} X_{kn} + \delta_{i+n} X_{km}, \quad (74)$$

where δ_q is one if $q = 0$, and zero otherwise. These operators correspond to the diagram B_l if we identify

$$X_{i-i} \equiv H_i, X_{\pm i \pm k} \equiv E_{(\pm e_i \pm e_k)}, X_{0 \pm k} \equiv E_{(\pm e_k)} \quad (i, k > 0) . \quad (75)$$

For the symplectic group in $2l$ dimensions, which leaves invariant the anti-symmetric bilinear form

$$\sum_{k=1}^{l} (x^k y^{-k} - x^{-k} y^k) ,$$

we may choose the infinitesimal operators

$$X_{ik} = X_{ki} = \varepsilon^i x^i \frac{\partial}{\partial x^{-k}} + \varepsilon^k x^k \frac{\partial}{\partial x^{-i}}, \quad (i, k = \pm 1, \ldots, \pm l), \quad (76)$$

with the commutation relations

$$[X_{ik} X_{mn}] = \varepsilon^m \delta_{k+m} X_{in} + \varepsilon^n \delta_{k+n} X_{im} + \varepsilon^m \delta_{i+m} X_{kn} + \varepsilon^n \delta_{i+n} X_{km}, \quad (77)$$

where ε^q is $+1$ if q is positive and -1 if q is negative. These operators correspond to the diagram C_l if we make the identification

$$X_{i-i} \equiv H_i, \quad X_{\pm i \pm k} \equiv E_{(\pm e_i \pm e_k)} \quad (i, k > 0). \quad (78)$$

For the orthogonal group in $2l$ dimensions, which leaves the quadratic form $\sum\limits_{k=1}^{l} x^k x^{-k}$ invariant, we may choose the same infinitesimal operators as in B_l, with the same commutation relations, except that now $i, k \neq 0$. These operators correspond to the diagram D_l if we make the same identification as in B_l.

Part III. The representations of the semi-simple groups

1. Representations and weights

A group of linear transformations of a vector space R which is homomorphic to a given group is called a *representation* of this group. The dimension, N, of R is called the *degree* of the representation. If s and t are two elements of the group, and $U(s)$ and $U(t)$ the corresponding matrices of the representation, then $U(s) U(t) = U(s t)$. Two representations $U(s)$ and $V(s)$ are called *equivalent* if there is a constant matrix A such that

$$A U(s) A^{-1} = V(s)$$

for every element s.

A representation is *reducible* if it leaves a subspace R_1 of R invariant. If this is the case, the matrices of the representation can be given the form

$$\begin{pmatrix} A_1 & B \\ 0 & A_2 \end{pmatrix}, \quad (79)$$

where A_1 is a matrix whose dimensions equal that of R_1. If the representation leaves invariant two subspaces R_1 and R_2 such that $R_1 + R_2 = R$, then the representation can be written as

$$\begin{pmatrix} A_1 & 0 \\ 0 & A_2 \end{pmatrix}. \quad (80)$$

We say in this case that the representation is *fully reducible*, or *decomposable*.

A Lie group is determined by the r infinitesimal operators and their commutation relations. Similarly, a representation of a Lie group is determined if we have r matrices, D_ϱ, which satisfy the equation

$$D_\varrho D_\sigma - D_\sigma D_\varrho \equiv [D_\varrho D_\sigma] = c_{\varrho\sigma}^\tau D_\tau . \tag{81}$$

In particular, we may ask for a standard representation with matrices H_i and E_α which satisfy the relations (50). These same letters, which denoted infinitesimal operators in the previous work, will in this lecture consistently be used for the corresponding matrices.

Let u be a vector in the space R such that

$$H_i u = m_i u \qquad (i = 1, \ldots, l) . \tag{82}$$

Thus, u is a simultaneous eigenvector of the l matrices H_i. The set of eigenvalues $m_1, \ldots, m_l$ are the covariant components of a vector in the l-dimensional space. We shall call this vector the *weight* of u; from now on the l-dimensional space will be called the *weight space*. Evidently, u is also the eigenvector of the matrix $\lambda^i H_i$ corresponding to the eigenvalue

$$(\lambda\, m) = \lambda^i m_i . \tag{83}$$

A weight will be called *simple* if to it belongs only one eigenvector.

The existence and various properties of the weights will now be proved.

A. Every representation has at least one weight.

Proof: H_1 has at least one eigenvalue, say m_1; let R_1 be the subspace of R spanned by the eigenvectors of H_1 belonging to m_1. Since $H_1 H_2 u = H_2 H_1 u = m_1 H_2 u$, it follows that $H_2 R_1 = R_1$. H_2 has at least one eigenvector in its invariant subspace R_1. Continuing the process, which is possible because every matrix has at least one eigenvector in every invariant subspace, we arrive at the subspace R_l which consists of the simultaneous eigenvectors of $H_1, \ldots, H_l$ corresponding to the weight $m \equiv (m_1, \ldots, m_l)$.

B. A vector u of weight m which is a linear combination of vectors u_k of weights $m^{(k)}$, all different from m, must vanish.

Proof: We form the matrix $\prod_k \lambda^i (H_i - m_i^{(k)})$ and let it operate on the equation $u = \sum_k u_k$. Since all H commute, each factor annihilates a term in the sum. Since the λ^i are arbitrary, the left hand side is also zero only if u vanishes.

C. From B it follows that vectors with different weights are linearly independent, so that there are at most N different weights.

D. If u is a vector of weight m, then $H_i u$ and $E_\alpha u$ have definite weights, m and $m + \alpha$ respectively.

Proof: For $H_i u$ this is an immediate consequence of (50). For $E_\alpha u$ we have

$$H_i E_\alpha u = [H_i E_\alpha] u + E_\alpha H_i u = (\alpha_i + m_i) E_\alpha u . \tag{84}$$

E. If the representation is irreducible the H_i may simultaneously be expressed in diagonal form.

Proof: Starting with a vector u having a definite weight, we consider the space R_1 spanned by all possible products

$$\ldots E_\gamma\, E_\beta\, E_\alpha\, u \,, \tag{85}$$

each of which, according to D, has a definite weight. Evidently, $E_\varrho R_1 = R_1$. Thus, since the representation is assumed irreducible, R_1 coincides with R, and the vectors (85) span R. If we select from them as a basis N linearly independent vectors, each is an eigenvector of all H_i, which thus have been diagonalized.

F. For any weight m and root α, $\dfrac{2(m\,\alpha)}{(\alpha\,\alpha)}$ is an integer and $m - \dfrac{2(m\,\alpha)}{(\alpha\,\alpha)}\,\alpha$ is a weight.

Proof: The proof is analogous to the proof of the Theorem of § 2, part II, except that the weights are, in general, not simple, while in the previous case Cartan's theorem enabled us to assume that all non-vanishing roots are simple. We shall point out only the differences in the proofs.

We start out from a vector u_0 of weight m such that $m + \alpha$ is not a weight, and form the series of vectors

$$u_1 = E_{-\alpha}\, u_0 \,, \quad u_2 = E_{-\alpha}\, u_1, \ldots \,.$$

The relation

$$E_\alpha\, u_{j+1} = \mu_{j+1}\, u_j \,, \tag{86}$$

which, because of the possible multiplicity of weights, is not as evident as its counterpart (54), may be proved by induction. Assume (86) to be true for a certain $j - 1$; then

$$E_\alpha\, u_{j+1} = E_\alpha E_{-\alpha}\, u_j = [E_\alpha E_{-\alpha}]\, u_j + E_{-\alpha} E_\alpha\, u_j = \alpha^i H_i\, u_j + \mu_j E_{-\alpha} u_{j-1}$$
$$= [(\alpha\, m) - j(\alpha\, \alpha)]\, u_j + \mu_j\, u_j \,.$$

Hence (86) is true for j if it is true for $j - 1$, and we have

$$\mu_{j+1} = (\alpha\, m) - j(\alpha\, \alpha) + \mu_j \,, \tag{87}$$

corresponding to (55). But $\mu_0 = 0$, by D, since $m + \alpha$ is not a root, and therefore (86) holds with $j + 1 = 0$ and $\mu_0 = 0$. The rest of the proof parallels that of the analogous theorem for the roots.

G. By projecting the space R moduli $u_0, \ldots, u_g$ in a space of $N - (g + 1)$ dimensions and by repeating the same considerations as in F, it may be proved that m and $m - \dfrac{2(\alpha\, m)}{(\alpha\, \alpha)}\, \alpha$ have the same multiplicity. All possible weights belong to a lattice which is invariant under the group S generated by the reflections with respect to the hyperplanes through the origin perpendicular to the roots. Weights which can be obtained from one another by operations of S are called *equivalent* and have the same multiplicity.

In the group A_l we have $(\alpha\, \alpha) = |e_i - e_k|^2 = 2$; hence a weight

$$m = m_1\, e_1 + m_2\, e_2 + \cdots + m_{l+1}\, e_{l+1} \tag{88}$$

must satisfy the condition that $2(m \cdot (e_i - e_k))/2 = m_i - m_k$ be an integer and, in addition, that

$$\sum_{i=1}^{l+1} m_i = 0 , \tag{88'}$$

which follows from (72). Therefore the m_i are fractions with denominator $l + 1$ which differ by integers. According to F and G, a weight equivalent to m is

$$m - (m_i - m_k) (e_i - e_k) = m_1 e_1 + \cdots + m_k e_i + \cdots + m_i e_k + \cdots + m_{l+1} e_{l+1} ;$$

hence the group S is the group of permutations of the components of m.

In B_l we have, in addition to the condition that $m_i - m_k$ be an integer, the further condition that $2(m \cdot e_i)$ be an integer. Therefore the components of any weight are either all integers or all half-integers. The group S is the group of permutations of the components with any number of changes of sign.

In C_l the additional condition is that $2(m \cdot 2e_i)/4$ be an integer, and therefore all components are integers. The group S is the same as in B_l.

In D_l we find that both $m_i - m_k$ and $m_i + m_k$ are integers. Therefore the weights are the same as in B_l, but the group S is only the group of permutations of the components with an even number of changes of sign.

2. The classification of the irreducible representations

We shall introduce a convention according to which the weights of the representations can be ordered. A weight $(m_1, \ldots, m_l)$ is said to be positive if the first non-vanishing component is positive. One weight is said to be *higher* than another if the difference between them is positive. A weight is called *dominant* if it is higher than its equivalents.

Theorem 1. If a representation is irreducible, its highest weight is simple.

Proof: Assume that the vector u_0 belongs to the highest weight, $m^{(0)}$. According to D of § 1 it is sufficient to prove that every vector of the form

$$\ldots E_\delta E_\gamma E_\beta E_\alpha u_0 \tag{89}$$

which is of weight $m^{(0)}$ can be written as $k u_0$, where k is a constant. We shall show, in addition, that k depends only on the series $\alpha, \beta, \gamma, \delta, \ldots$, and on the weight $m^{(0)}$. It is clear from D that $\cdots + \delta + \gamma + \beta + \alpha = 0$. Therefore at least one of the roots must be positive. Let us say that γ is the first positive root (from the right). Replacing $E_\gamma E_\beta$ by $E_\beta E_\gamma + [E_\gamma E_\beta]$, and so on until E_γ acts directly on u_0, and remembering that $E_\gamma u_0 = 0$, we obtain a sum of terms with fewer matrices E than (89) but still of weight $m^{(0)}$. Continuing this process until there are no more operators of positive weight, we arrive at a sum of products of H_i acting on u_0, and these are finally converted into a polynomial of the components of $m^{(0)}$ multiplying u_0. The coefficients in this polynomial depend, evidently, only on the set of roots $\alpha, \beta, \gamma, \delta, \ldots$, and not on the particular representation.

Theorem 2. Two irreducible representations are equivalent if their highest weights are equal.

Proof: We distinguish the two representations D and D' by using unprimed quantities for D and primed ones for D'. Let u_0 and u_0' be the vectors of the highest weight $m^{(0)}$, which is assumed to be the same for both D and D', and construct all possible vectors $u_j = \cdots = E_\gamma E_\beta E_\alpha u_0$ and correspondingly $u_j' = \cdots = E_\gamma' E_\beta' E_\alpha' u_0'$. It was shown in D of § 1 that these vectors span the whole space and that each has a definite weight. The equivalence of the two representations will be proved if we show that to any linear relation which exists between the unprimed vectors there corresponds a linear relation with the same coefficients between the corresponding primed vectors. Assume there is a relation

$$\gamma_1 u_1 + \gamma_2 u_2 + \cdots = 0; \tag{90}$$

then, using the same coefficients γ we can construct a vector

$$\gamma_1 u_1' + \gamma_2 u_2' + \cdots = w'. \tag{90'}$$

The vectors w' for all possible relations (90′) form a subspace R_1' of R', and it is easily seen that R_1' is an invariant subspace under the operations of the group. Since D' is irreducible we must have $R_1' = 0$ unless R_1' consists of the whole space R'. The last alternative is excluded since u_0' is certainly not in R_1'. For if $w' = u_0'$, according to C of § 1, the left-hand side of (90) contains only vectors of weight $m^{(0)}$. Theorem 1 would then lead to a relation

$$\gamma_1 k_1 + \gamma_2 k_2 + \cdots \neq 0$$

which however, is incompatible with the corresponding relation

$$\gamma_1 k_1 + \gamma_2 k_2 + \cdots = 0$$

derived from (90), the k being the same for the two representations.

The connection between highest weights and irreducible representations is completed when we show that there exists an irreducible representation which has any dominant weight as its highest weight. Indeed, CARTAN has proved that

(A) For every simple group of rank l there are l *fundamental* dominant weights $L^{(1)}, \ldots, L^{(l)}$ such that if a dominant weight L is given, it is a linear combination

$$L = \sum_{i=1}^{l} \varkappa_i L^{(i)} \tag{91}$$

with non-negative integral coefficients;

(B) There exist l fundamental irreducible representations $g_1, g_2, \ldots, g_l$ which have the fundamental weights as their highest weights.

Since it is easy to see that the weights of the Kronecker product $A \times B$ of two representations are all the sums of one weight of A and one weight of B, the Kronecker product representation

$$G = g_1 \times g_1 \times \cdots \times g_2 \times g_2 \cdots \times \cdots \tag{92}$$
$$\varkappa_1 \text{ times} \qquad \varkappa_2 \text{ times}$$

has as highest weight exactly the weight L. G will, in general, be reducible, but one of its irreducible constituents will have L as its highest weight.

CARTAN proved (A) and (B) for every simple group separately[*]. We shall here sketch as an example the proofs for the groups A_l and B_l.

A_l. The components of a dominant weight satisfy the relation $L_1 \geq L_2 \geq \cdots \geq L_{l+1}$. If we assume as fundamental weights

$$L^{(1)}: \frac{l}{l+1}, \quad -\frac{1}{l+1}, \ldots, -\frac{1}{l+1}$$

$$L^{(2)}: \frac{l-1}{l+1}, \quad \frac{l-1}{l+1}, \quad -\frac{2}{l+1}, \ldots, -\frac{2}{l+1} \tag{93}$$

$$\cdots\cdots\cdots\cdots\cdots\cdots\cdots$$

$$L^{(l)}: \frac{1}{l+1}, \quad \frac{1}{l+1}, \ldots, \frac{1}{l+1}, \quad \frac{-l}{l+1}$$

it can be verified that (91) is satisfied by setting

$$\varkappa_i = L_i - L_{i+1}. \tag{94}$$

The fundamental representations corresponding to the highest weights (93) are the linear unimodular group in $l+1$ dimensions itself and the transformations induced by this group on the antisymmetric tensors of rank $2, 3, \ldots, l$.

It may be shown that a tensor of rank f in the $l+1$ dimensional space which has the symmetry defined by the partition $(f_1, f_2, \ldots, f_{l+1})$ with $\sum_{i=1}^{l+1} f_i = f$ is a basis of the representation whose highest weight has the components $L_i = f_i - \frac{f}{l+1}$.

B_l. The components of a dominant weight satisfy the relation $L_1 \geq L_2 \geq \cdots \geq L_l \geq 0$. If we take as fundamental weights

$$L^{(1)}: \frac{1}{2} \frac{1}{2} \cdots \frac{1}{2}$$

$$L^{(2)}: 1\, 0\, 0 \ldots 0$$

$$L^{(3)}: 1\, 1\, 0\, 0 \ldots 0 \tag{95}$$

$$\cdots\cdots\cdots$$

$$L^{(l)}: 1\, 1\, 1 \ldots 1\, 0$$

it is easy to see that (91) is satisfied by setting

$$\varkappa_1 = 2L_l$$

$$\varkappa_i = L_{i-1} - L_i \qquad (i > 1). \tag{96}$$

The fundamental representations corresponding to the highest weights (95) are the double-valued representation of degree 2^l, the orthogonal group in $2l+1$ dimensions, and the transformations induced by this group on the antisymmetric tensors of rank $2, 3, \ldots, l-1$.

---

[*] CHEVALLEY, Compt. Rend. **227**, 1136 (1948) has given a proof of the whole theorem which does not make use of the particular structure of the different groups.

It may be shown that a tensor of rank f with vanishing trace in the $2l + 1$ dimensional space which has a symmetry defined by the partition $(f_1, \ldots, f_l, 0, \ldots, 0)$ is the basis of the representation whose highest weight has components $L_i = f_i$.

3. The problem of full reducibility

Having classified the irreducible representations of a group, we are in a position to classify all its representations if we know that every reducible representation is fully reducible, i. e. decomposable into its irreducible constituents.

It is well known that the representations of finite groups are fully reducible, and that the proof of this is based on the possibility of summing over all elements of a group representation. For continuous groups the analog of this summation is an integration for which, however, the question of convergence arises. WEYL has proved that if we impose some particular reality condition (this is called the *unitary restriction*) on the coefficients e^ϱ of the general infinitesimal element $e^\varrho X_\varrho$ of a semi-simple group, the group is restricted to a subgroup for which the integrations converge and full reducibility may be proved. It follows from the full reducibility of any infinitesimal representation $D_1, \ldots, D_r$ that the general element $e^\varrho D_\varrho$ is fully reducible even if the e^ϱ no longer obey the unitary restriction. The representations of every semi-simple group are therefore fully reducible.

Under the unitary restriction the linear group becomes the unitary group, and the orthogonal group becomes that of real rotations. WEYL's proof involves integration over the entire group. A purely infinitesimal proof of the full reducibility was given by CASIMIR for the three-dimensional orthogonal group O_3. He considered the operator[*]

$$G = J_x^2 + J_y^2 + J_z^2 \tag{97}$$

which is known to commute with J_x, J_y, and J_z. If the representation is irreducible, then Schur's lemma states that G is of the form

$$G = \lambda \cdot 1 , \tag{98}$$

where

$$\lambda = j(j + 1) \ \left(j = 0, \frac{1}{2}, 1, \frac{3}{2}, \ldots\right) . \tag{98'}$$

If the representation is reducible, and has for example two irreducible constituents, the infinitesimal operators may be brought to the form (79), so that G can be written

$$G = \begin{pmatrix} \lambda 1 & K \\ 0 & \lambda' 1 \end{pmatrix} . \tag{99}$$

---

[*] Hereafter, to avoid confusion, we shall use J_x, J_y, J_z instead of L_1, L_2, L_3 (see p. 39).

If $\lambda \neq \lambda'$, then by application of the transformation

$$T = \begin{pmatrix} 1 & \dfrac{K}{\lambda - \lambda'} \\ 0 & 1 \end{pmatrix} \qquad (100)$$

we obtain

$$T G T^{-1} = \begin{pmatrix} \lambda 1 & 0 \\ 0 & \lambda' 1 \end{pmatrix}$$

The same transformation also decomposes J_x, J_y, and J_z, since they commute with G. The decomposition fails if $\lambda = \lambda'$, but in this case the two irreducible constituents of the representation are equivalent and full reducibility may be proved by quite simple considerations. We shall see in the next section how this proof may be generalized so as to apply to any semi-simple group.

4. Casimir's operator and its generalization

We have seen in § 2 that every irreducible representation is characterized by its highest weight $L \equiv (L_1, \ldots, L_l)$. But in the group O_3, j is not only the highest value of m, i. e. the highest eigenvalue of J_z for the given representation, but is also connected with the eigenvalues of G, which are common to the whole basis of an irreducible representation. The connection is one-to-one, since it follows from (98′) that

$$j = \pm \sqrt{\lambda + \frac{1}{4}} - \frac{1}{2},$$

but only the upper sign gives a j which is a dominant weight.

The generalization of G for any semi-simple group was given by CASIMIR, who introduced the operator

$$G = g^{\varrho\sigma} X_\varrho X_\sigma, \qquad (101)$$

which commutes with every X_τ

$$[G\, X_\tau] = g^{\varrho\sigma} X_\varrho [X_\sigma X_\tau] + g^{\varrho\sigma} [X_\varrho X_\tau]\, X_\sigma$$
$$= (c_\tau^{\varrho\lambda} + c_\tau^{\lambda\varrho})\, X_\varrho X_\lambda = 0$$

by the antisymmetry of the structure constants. The eigenvalues of G may be calculated if we use the standard basis and write

$$G = g^{ik} H_i H_k + \sum_\alpha E_\alpha E_{-\alpha}. \qquad (102)$$

Let L be the highest weight of an irreducible representation and u be a vector of this weight in the space R. Then $E_\alpha u = 0$ for positive roots α, and

$$G u = g^{ik} L_i L_k u + \sum_{\alpha^+} [E_\alpha E_{-\alpha}]\, u = [(L\, L) + \sum_{\alpha^+} (\alpha\, L)]\, u \qquad (103)$$

where $\sum\limits_{\alpha^+}$ denotes summation over positive roots only. By introducing the vectors

$$R = \frac{1}{2} \sum_{\alpha^+} \alpha \tag{104}$$

and

$$K = L + R , \tag{105}$$

we can write for the eigenvalue of G

$$\lambda = L^2 + 2(R\,L) = K^2 - R^2 . \tag{106}$$

It is easy to see that while a highest weight determines an eigenvalue of the Casimir operator, the converse is not generally true, and the fact is not surprising as we cannot expect that the single number λ is sufficient to determine l numbers L_i.

CASIMIR used the operator G in order to extend to any semi-simple group his proof of full reducibility, but was unable to apply it to the cases where inequivalent representations belong to the same eigenvalue of G. The latter case was treated by VAN DER WAERDEN by the use of considerations entirely foreign to CASIMIR's original approach.

Another way of doing it is to generalize Casimir's operator by constructing a complete set of operators which commute with every operator of the group and whose eigenvalues characterize the irreducible representations.

A possible generalization of G is provided by the operators

$$\gamma_{\alpha_1 \alpha_2 \ldots \alpha_n} X^{\alpha_1} X^{\alpha_2} \ldots X^{\alpha_n}$$

with

$$\gamma_{\alpha_1 \alpha_2 \alpha_n \ldots} = c^{\beta_2}_{\alpha_1 \beta_1} c^{\beta_3}_{\alpha_2 \beta_2} \cdots c^{\beta_1}_{\alpha_n \beta_n} ,$$

and it is easy to verify that each of these operators commutes with every X_ϱ. But these still do not suffice, since it is found for example that for irreducible representations contragredient to each other and inequivalent, they have the same eigenvalues.

We therefore examine the conditions imposed on a general function of the infinitesimal operators, $F(X^\varrho)$, by the requirement that it commute with every operator of the group

$$[X_\sigma F] = 0 . \tag{107}$$

It is well known that this expression can be written as

$$[X_\sigma X^\tau] \frac{\partial F}{\partial X^\tau} = c^{\tau\lambda}_\sigma X_\lambda \frac{\partial F}{\partial X^\tau} = c^\tau_{\lambda\sigma} X^\lambda \frac{\partial F}{\partial X^\tau} ,$$

where the products $X^\lambda \partial F / \partial X^\tau$ are suitably ordered. Comparison of this expression with (19) shows that the functions satisfying (107) may be constructed from the invariants of the adjoint group, which are characterized by

$$E_\sigma F(e^\varrho) = 0 , \tag{108}$$

by substituting e^ϱ for X^ϱ and ordering the terms.

By applying an operator F which satisfies (107) to any vector of the space R of an irreducible representation we obtain, according to Schur's lemma, $F u = \lambda u$, where λ is independent of the particular choice of u. If the vector of highest weight is chosen, we find from § 2, Theorem 1, that

$$\lambda = \varphi(L_1; L_2, \ldots, L_l) \equiv \varphi(L) . \tag{109}$$

In order to characterize the representation we need l operators of this kind such that the system of equations

$$\lambda_i = \varphi_i(L) \qquad (i = 1, \ldots, l) \tag{109'}$$

has not more than one solution L which is a dominant weight. To prove the existence of such a set of operators it suffices to prove that

A) If we express the λ_i as functions of K instead of L, the functions

$$\lambda_i = f_i(K) \tag{110}$$

are invariant under the transformations of the group (S) defined on page 47.

B) For any simple (or semi-simple) group there exists a set of l (polynomial) invariants of the adjoint group such that the product of the degrees of these polynomials equals the order of (S).

According to A), the system (110) has, together with a solution K, any solution SK [which means the vector obtained from K by an operation S of the group (S)], and according to B), the number of solutions exactly equals the number of vectors SK ⋆; hence the system (110) has only one solution which is a dominant vector. Also (109′) has only one solution L which is a dominant vector, because if a solution K of (110) is not dominant and is lower, say, than SK, then also $K - R$ is not dominant, since

$$K - R < SK - R < SK - SR = S(K - R) .$$

We shall prove A) by making use of the properties of the whole group, since it has not been possible so far to construct a proof which uses only the infinitesimal group. In any representation, (5) reads

$$U(\delta a) U(a) = U(a + \mathrm{d}\, a) .$$

As the D_ϱ of (81) are the infinitesimal elements of the representation we may write this as

$$\sum_t (q \,|1 + \delta a^\varrho D_\varrho| \, t) \, (t \,|U(a^\sigma)| \, s) = (q \,|U(a^\sigma + \mu^\sigma_\varrho \, \delta a^\varrho)| \, s) \quad \text{by (8)}$$

$$= (q \,|U(a^\sigma)| \, s) + \mu^\sigma_\varrho \, \delta a^\varrho \frac{\partial}{\partial a^\sigma} \, (q \,|U(a^\sigma)| \, s) . \; ⋆⋆$$

⋆ From the definition of R it follows easily that no SK coincides with K, and therefore no two SK coincide.

⋆⋆ No summation over σ (ed.).

Comparison with (15) shows that

$$A_\varrho(q \,|U(a^\sigma)|\, s) = \sum_t (q \,|D_\varrho|\, t) \, (t \,|U(a^\sigma)|\, s) \,,$$

where A_ϱ is the infinitesimal operator of the first parameter group. Consequently for any function $f(X_\varrho)$ we have

$$f(A_\varrho) \, (q \,|U(a^\sigma)|\, s) = \sum_t (q \,|f(D_\varrho)|\, t) \, (t \,|U(a^\sigma)|\, s) \,. \tag{111}$$

In particular, if $f(X_\varrho)$ is $F(X_\varrho)$ satisfying (107), then $F(D_\varrho)$ is diagonal according to Schur's lemma, and we get

$$F(A_\varrho) \, (q \,|U(a^\sigma)|\, s) = \lambda (q \,|U(a^\sigma)|\, s) \,, \tag{112}$$

so that each matrix element of the representation is an eigenfunction of $F(A_\varrho)$. It follows that the trace of the matrix $(q \,|U(a^\sigma)|\, s)$, which is called the *character*, χ, of the representation, is also an eigenfunction corresponding to the same eigenvalue. Since the trace of a matrix is invariant under similarity transformations, it follows that the character is not a function of the individual elements of the group, but rather of the classes of conjugate elements. The classes of a semi-simple group of rank l depend on l parameters; by choosing them as a suitable set $\varphi^1, \ldots, \varphi^l$, WEYL [6] has given a general formula,

$$\chi(L; \varphi) = \frac{\xi(K)}{\xi(R)} \tag{113}$$

for the characters of a unitary restricted semi-simple group, where K is defined by (105) and

$$\xi(K) = \sum_S \delta_S \, e^{i(SK)_j \, \varphi^j} \,; \tag{114}$$

δ_S is plus or minus one depending on the parity of the element S.

If we now apply to K an operation S, the character is left invariant except for a possible change of sign; hence the eigenvalue (110) to which the character belongs as eigenfunction is invariant under the operations of (S). Q. E. D.

As an example consider the group of rotations in three dimensions, R_3. Any element of this group can be obtained by a similarity transformation from the diagonal matrix with elements $e^{im\varphi}$ $(-l \le m \le l)$ where φ is an angle of rotation around a properly chosen axis. Thus, φ is a function of the class and the character of R_3' is

$$\sum_{m=-l}^{l} e^{im\varphi} = \frac{e^{i\left(l+\frac{1}{2}\right)\varphi} - e^{-i\left(l+\frac{1}{2}\right)\varphi}}{e^{i\frac{\varphi}{2}} - e^{-i\frac{\varphi}{2}}} \,, \tag{115}$$

which is the value given by (114) where $k = l + \dfrac{1}{2}$.

In order to construct invariants of the adjoint group we construct the determinant

$$\Delta = \det (q \,|e^\varrho D_\varrho - \omega|\, s) = \det a_{qs} \,, \tag{116}$$

where ω is an arbitrary number and $(q\,|D_\varrho|\,s)$ an arbitrary representation. The determinant is an invariant of the adjoint group, for

$$E_\sigma \Delta = e^\varrho\, c_{\varrho\sigma}^\tau\, \frac{\partial \Delta}{\partial e^\tau} = e^\varrho\, c_{\varrho\sigma}^\tau \sum_{qs} \frac{\partial \Delta}{\partial a_{qs}}\, \frac{\partial a_{qs}}{\partial e^\tau}$$

$$= e^\varrho\, c_{\varrho\sigma}^\tau \sum_{qs} \frac{\partial \Delta}{\partial a_{qs}}\, (q\,|D_\tau|\,s)$$

$$= e^\varrho \sum_{qst} \frac{\partial \Delta}{\partial a_{qs}}\, [(q\,|D_\varrho|\,t)\,(t\,|D_\sigma|\,s) - (q\,|D_\sigma|\,t)\,(t\,|D_\varrho|\,s)] \quad \text{by (81)}$$

$$= \sum_{qst} \frac{\partial \Delta}{\partial a_{qs}}\, [(a_{qt} + \omega\,\delta_{qt})\,(t\,|D_\sigma|\,s) - (q\,|D_\lambda|\,t)\,(a_{ts} + \omega\,\delta_{ts})]$$

$$= \sum_{qst} \frac{\partial \Delta}{\partial a_{qs}}\, [a_{qt}(t\,|D_\sigma|\,s) - (q\,|D_\sigma|\,t)\,a_{ts}]$$

$$= \sum_{st} \Delta\, \delta_{st}(t\,|D_\sigma|\,s) - \sum_{qt} \Delta\, \delta_{qt}(q\,|D_\sigma|\,t) = 0\,.$$

Δ is a polynomial in ω, and evidently the coefficient of each power of ω is separately an invariant. That this method yields a set of invariants which satisfy the conditions stated in B) is shown separately for each simple group in reference [14].

In conclusion, we can now state that for every semi-simple group there exists a set of l functions $F_i(X_\varrho)$ which commute with every operator of the group and whose eigenvalues characterize the irreducible representations. They constitute the extension to every semi-simple group of the operator (97) for the three-dimensional rotation group.

5. Miscellaneous problems

Finally, we know a number of general properties of the irreducible representation of O_3, and we want to see to what extent they may be generalized to all semi-simple groups.

1. The dimension of an irreducible representation is $2j + 1$ for O_3. By calculating the value of (113) for the identity element, WEYL has found that the dimension of any irreducible representation is given by

$$\prod_{\alpha^+} \frac{(\alpha K)}{(\alpha R)}\,. \tag{117}$$

2. In O_3 the eigenvalues of J_z are non-degenerate and hence suffice to label the basis of the representations. The natural extension of the eigenvalues of J_z are the weights, but in general they are not simple. If γ_m is the multiplicity of the weight m, WEYL has shown that the character of the representation has the form

$$\chi = \sum_m \gamma_m\, e^{im_j\varphi^j}\,, \tag{118}$$

so that the coefficients of the Fourier expansion of expression (113) give the multiplicities.

If the multiplicity is different from unity we need some additional operators $k(X_\varrho)$, all commuting with each other and with H_i, whose eigenvalues will enable us to distinguish the different eigenvectors of a given weight; we must first find out how many such operators will be needed.

If the basis is chosen so that not only H_i but also $k(X_\varrho)$ are diagonal, then by setting $f(X_\varrho) = k(X_\varrho)$ in (111) we obtain

$$k(A_\varrho) \, (q \, |U(a^\sigma)| \, s) = k_q(q \, |U(a^\sigma)| \, s) \tag{119}$$

where k_q is the eigenvalue of $k(X_\varrho)$ corresponding to the row q; then $(q \, |U(a^\sigma)| \, s)$ is an eigenfunction of $k(A_\varrho)$ corresponding to this eigenvalue. Similarly by considering the second parameter group it may be shown that $(q \, |U(a^\sigma)| \, s)$ is also an eigenfunction of $k(B_\varrho)$ corresponding to the eigenvalue k_s.

In order to identify the functions $(q \, |U(a^\sigma)| \, s)$ of the r parameters completely, we need a set of at least r commuting operators acting on these parameters.

We are already in possession of the l commuting operators $F_i(A_\varrho)$ $= F_i(B_\varrho)$. Hence we still need $\dfrac{r-l}{2}$ * operators $k(X_\varrho)$, in order to have $\dfrac{r-l}{2}$ operators $k(A_\varrho)$ and the same number of $k(B_\varrho)$. However, l such operators $k(X_\varrho)$ are already known to us; they are the H_i themselves. Hence for the set of commuting operators to be complete we need at least to construct $\dfrac{r-3l}{2}$ operators $k(X_\varrho)$.

In the particular case of the group O_3, $\dfrac{r-3l}{2} = 0$, and it is well known that the operators J_z and J^2 form the complete set. The problem of finding the complete set of operators $k(X_\varrho)$ has so far been solved only for some types of simple groups.

3. Explicit construction of the irreducible representations. The representations of the infinitesimal operators of O_3 are the diagonal matrix J_z and the matrices $J_x \pm i J_y$ whose only non-vanishing matrix elements are given by

$$(j \, m \pm 1 \, |J_x \pm i J_y| \, j \, m) = \sqrt{(j \pm m + 1) \, (j \mp m)} \; . \tag{120}$$

In the general case, the corresponding formula should be

$$(L M + \alpha \, k_q^{(h)} \, |E_\alpha| \, L M \, k_s^{(h)}) = f(L M \, \alpha \, k_q^{(h)} \, k_s^{(h)}) \, , \tag{121}$$

but as long as the $k^{(h)}$ are not known it is impossible to give an explicit form to the function f. Later on we shall present some special methods for solving this problem in particular cases in which we are interested.

4. Decomposition of the Kronecker product:

In O_3 this is done by the *Clebsch-Gordan series*

$$\mathscr{D}(J_1) \times \mathscr{D}(J_2) = \sum_{J=|J_1-J_2|}^{J_1+J_2} \mathscr{D}(J) \; . \tag{122}$$

---

* The structure of semi-simple groups assures that this number is always an integer.

In general we have seen that

$$\mathscr{D}(L^{(1)}) \times \mathscr{D}(L^{(2)}) = \mathscr{D}(L^{(1)} + L^{(2)}) + \cdots \tag{123}$$

but we were not in a position to say anything about the other terms of the series. The coefficients in this series have been given by BRAUER and WEYL[*] by using the characters of the representations.

But this is only the first part of the problem, since we not only need to know which irreducible representations are contained in a Kronecker product, but we also want to calculate the matrix which actually decomposes the Kronecker product.

For O_3 this problem has been solved in several different ways.

The classical Clebsch-Gordan method exploits the homomorphism between O_3 and the unimodular group in two dimensions (which is the basis of spinor calculus), but this method is applicable only to this particular case and is not capable of generalization. WIGNER solved the same problem by performing integrations over the whole group, but actually it is sufficient to consider the infinitesimal representation, as will be indicated here.

The transformation coefficients $(j_1\, m_1\, j_2\, m_2 | j_1\, j_2\, J\, M) \equiv (m_1\, m_2 | J\, M)$ are defined by the relation

$$\sum_{m_1'' m_2''} (J''M'' \,|\, m_1'' m_2'') \{(m_1'' |j_{1x} \pm i j_{1y}| m_1') + (m_2'' |j_{2x} \pm i j_{2y}| m_2')\} \cdot$$
$$\cdot (m_1'\, m_2' \,|\, JM) = (J''M'' \,|\, J_x \pm i J_y| \,JM)\, \delta_{JJ''}.$$

Matrix multiplication of this equation from the left by $(m_1\, m_2 | J''M'')$ gives

$$\sum_{m_1' m_2'} \{(m_1 |j_{1x} \pm i j_{1y}| m_1') + (m_2 |j_{2x} \pm i j_{2y}| m_2')\}\, (m_1'\, m_2' \,|\, JM)$$
$$= \sum_{M''} (m_1 m_2 \,|\, JM'')\, (JM'' |J_x \pm i J_y| \,JM)$$

which, using (120), becomes the recursion formula

$$(m_1|j_{1x} \pm i j_{1y}| m_1 \mp 1)\, (m_1 \mp 1\, m_2 \,|\, JM) + (m_2|j_{2x} \pm i j_{2y}| m_2 \mp 1) \cdot$$
$$\cdot (m_1\, m_2 \mp 1 \,|\, JM) = (JM \pm 1 |J_x \pm i J_y| \,JM)\,(m_1\, m_2 \,|\, JM \pm 1). \tag{124}$$

If we take the upper sign and set $M = J$, we see that the right hand side vanishes and we find a set of equations which determines the different $(m_1\, m_2 | J\, J)$ apart from a common factor whose absolute value is fixed by normalization and whose phase is fixed by the convention that $(j_1\, J - j_1 | J\, J)$ be real and positive. Taking now the lower sign we obtain $(m_1 m_2 | J M - 1)$ from $(m_1\, m_2 | J M)$; hence by a "ladder" procedure starting from $M = J$ we get all the transformation coefficients.

This method would probably be the one best suited for extension to the other groups provided the right hand side of (121) were known explicitly.

---

[*] Reference [13], p. 229.

6. The full linear group and the unitary group

We saw on p. 44 that the full linear group in k dimensions (as well as its unitary subgroup) is not semi-simple. But, since it is the direct product of a semi-simple group with an Abelian group, the full linear group shares many properties with the semi-simple groups, including the possibility of bringing the commutation relations to the standard form (50) and all the results of §§ 1 and 2. It is also clear from p. 44 that, as the unimodular condition is omitted, H_i has now to be identified with X_{ii} and not with X'_{ii} defined by (69'); the components of the weights are now always integers, and the relations (72) and (88') which were obtained for the unimodular group do not hold for the full linear group.

It can be shown that a tensor of rank f in the k-dimensional space which has the symmetry defined by the partition $\Sigma \equiv (f_1, f_2, \ldots, f_k)$, with

$$f = f_1 + f_2 + \cdots + f_k$$

is a basis of the representation whose highest weight has the components f_i.

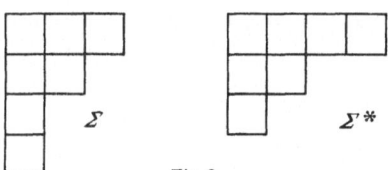

Fig. 2

We may conveniently illustrate a partition Σ by a *Young diagram* such as the one at the left, consisting of f boxes in k rows, the ith row containing f_i boxes.

A partition Σ^* is said to be *dual* to Σ if its diagram is obtained by interchanging the rows and columns of Σ.

In particular, the partition

$$f = \underbrace{1 + 1 + \cdots + 1}_{f} + \underbrace{0 + 0 + \cdots + 0}_{k-f}$$

which is possible if $k \geq f$, characterizes a representation $\mathscr{A}_f$ of degree $\binom{k}{f}$ whose basis is formed by the totally antisymmetrical tensors of rank f.

The irreducible representations of the full linear group do not decompose if we restrict the group to its unimodular subgroup, but the representations which belong to the partitions $(f_1, \ldots, f_k)$ and $(f_1 + e, f_2 + e, \ldots, f_k + e)$ become equivalent.

All the properties which we have stated for the full linear group hold also for its unitary subgroup.

Part IV. The eigenfunctions of the nuclear shells

1. Introduction

If we wish to calculate the energy levels of a system of many particles, the fact that we cannot solve directly the Schrödinger equation for the manybody problem forces us to proceed by successive approximations.

In atomic spectroscopy we assume that in the "zeroth approximation" every electron moves independently of the others in a central field which is the superposition of the fields of the nucleus and of the mean field produced by the other electrons. In this approximation we may assign to every electron four quantum numbers $n \, l \, m_s \, m_l$; as the zeroth order energy depends only on n and l, the electrons appear to be distributed in different *shells*, each characterized by a pair of values $n \, l$. Such a distribution is called a *configuration*.

The next step is to take as a perturbation the interaction between electrons in shells which are not closed, neglecting in first approximation the matrix elements which connect different configurations.

It is well known that applied to atomic spectroscopy this method gives good results. It is also well known that the theoretical arguments for using this method in nuclear spectroscopy are very weak, but that there is on the other hand some empirical evidence that the nucleons also are ordered in shells. We shall not, however, discuss here the validity of the nuclear shell model.

It is the purpose of the remaining lectures to show some applications of group theory to the classification of the levels of a nuclear shell and to the calculation in first approximation of the perturbation energy.

As is customary in dealing with problems of spectroscopy we shall use the standard notation of reference [15].

2. The coefficients of fractional parentage

If a shell contains one particle, the quantum numbers $m_\tau \, m_s \, m_l$ describe the state completely. If a shell contains two particles, one can use the quantum numbers $m_\tau^{(1)} \, m_s^{(1)} \, m_l^{(1)} \, m_\tau^{(2)} \, m_s^{(2)} \, m_l^{(2)}$ or, alternatively, $T \, S \, L \, M_T \, M_S \, M_L$ of which the second scheme is the more useful, since it diagonalizes the energy; the transformation leading from the one of these schemes to the other is given by the Clebsch-Gordan coefficients. A further advantage of the second scheme is that in it the states are either symmetrical or antisymmetrical, depending on the parity of $T + S + L$; the exclusion principle simply removes the states for which $T + S + L$ is even, without changing the scheme.

If we add to the allowed states of the configuration l^2 a third l-particle, we obtain a set of wave functions

$$\psi\left(l^2\left(T^{(12)} \, S^{(12)} \, L^{(12)}\right) l, \, T \, S \, L \, M_T \, M_S \, M_L\right), \qquad (125)$$

which are in general antisymmetrical only with respect to the first two particles, but not with respect to the third. If to (125) we apply the transformation

$$\psi\left(l^2\left(T^{(12)} \, S^{(12)} \, L^{(12)}\right) l, \, T \, S \, L \, M_T \, M_S \, M_L\right)$$

$$= \sum_{T^{(23)} \, S^{(23)} \, L^{(23)}} \psi\left(l, \, l \, l\left(T^{(23)} \, S^{(23)} \, L^{(23)}\right), \, T \, S \, L \, M_T \, M_S \, M_L\right) \times$$

$$\times \left(\frac{1}{2}, \frac{1}{2} \, \frac{1}{2} \, (T^{(23)}), \, T \, \middle| \, \frac{1}{2} \, \frac{1}{2} \, (T^{(12)}) \, \frac{1}{2}, \, T\right) \times \qquad (126)$$

$$\times \left(\frac{1}{2}, \frac{1}{2} \, \frac{1}{2} \, (S^{(23)}), \, S \, \middle| \, \frac{1}{2} \, \frac{1}{2} \, (S^{(12)}) \, \frac{1}{2}, \, S\right) \times$$

$$\times \left(l, \, l \, l \, (L^{(23)}), \, L \, \middle| \, l \, l \, (L^{(12)}) \, l, \, L\right),$$

we see that in the expansion appear some terms which will be symmetrical rather than antisymmetrical in the last two particles.

The eigenfunctions of the configuration l^3, which have to be anti-symmetrical in all three particles, span a subspace of the space spanned by the functions (125) and will thus be linear combinations of them

$$\psi(l^3 \alpha T S L) = \sum_{T^{(12)} S^{(12)} L^{(12)}} \psi(l^2(T^{(12)} S^{(12)} L^{(12)}) l, T S L) \times$$
$$\times (l^2(T^{(12)} S^{(12)} L^{(12)}), l, T S L |\} l^3 \alpha T S L) . \tag{127}$$

We have omitted $M_T M_S M_L$ from the notation because they play no role in the transformation; α distinguishes independent states of l^3 which have the same values of $T S L$. The notation $(|\})$ is a reminder that this trans-formation matrix is not square, since on the left side we have all states which are antisymmetrical in $(1, 2)$, while on the right we have only those states which are antisymmetrical in $(1, 2, 3)$. The coefficients of this linear combination are called *coefficients of fractional parentage* or, for short, c. f. p.

If (127) is to be antisymmetrical in all three particles, this requires that when (126) is substituted into (127), all those coefficients which belong to forbidden wave functions shall vanish, and it is easy to see that the necessary and sufficient condition for this is

$$\sum_{T^{(12)} S^{(12)} L^{(12)}} \left(\frac{1}{2}, \frac{1}{2} \frac{1}{2} (T^{(23)}), T \left| \frac{1}{2} \frac{1}{2} (T^{(12)}) \frac{1}{2}, T\right.\right) \times$$
$$\times \left(\frac{1}{2}, \frac{1}{2} \frac{1}{2} (S^{(23)}), S \left| \frac{1}{2} \frac{1}{2} (S^{(12)}) \frac{1}{2}, S\right.\right) (l, ll(L^{(23)}), L \,|\, ll(L^{(12)}) l, L) \times \quad (128)$$
$$\times (l^2(T^{(12)} S^{(12)} L^{(12)}) l, T S L |\} l^3 \alpha T S L) = 0$$

when $T^{(23)} + S^{(23)} + L^{(23)}$ is even.

This system of equations contains all the information we need for the configuration l^3, since the number of independent solutions for given $T S L$, which are distinguished by the parameter α, is the number of allowed states of this kind, and since we can also use the c. f. p. to calculate the interaction energy for these three-particle states

$$(l^3 \alpha T S L |E| l^3 \alpha' T S L) = 3 \sum_{T^{(12)} S^{(12)} L^{(12)}} (l^3 \alpha T S L \{| l^2(T^{(12)} S^{(12)} L^{(12)}) l,$$
$$T S L) E(T^{(12)} S^{(12)} L^{(12)}) \cdot (l^2(T^{(12)} S^{(12)} L^{(12)}) l, T S L |\} l^3 \alpha' T S L) ; \tag{129}$$

$E(T^{(12)} S^{(12)} L^{(12)})$ is the interaction energy for the two-particle system, and the factor 3 enters because there are three pairs of particles in the configuration.

The extension of these methods to a shell which contains n particles is in principle immediate. We start from a shell with $n - 1$ particles, for which we suppose the c. f. p. to be calculated. Then

$$\psi(l^n \alpha T S L) = \sum_{\alpha_1 T_1 S_1 L_1} \psi(l^{n-1}(\alpha_1 T_1 S_1 L_1) l, T S L) \times$$
$$\times (l^{n-1}(\alpha_1 T_1 S_1 L_1) l, T S L |\} l^n \alpha T S L) \tag{130}$$

where, analogously to (128), the c. f. p. satisfy the system of equations

$$\sum_{\alpha_1 T_1 S_1 L_1} \left(T_2, \tfrac{1}{2}\tfrac{1}{2}(T'), T \middle| T_2 \tfrac{1}{2}(T_1)\tfrac{1}{2}, T\right)\left(S_2, \tfrac{1}{2}\tfrac{1}{2}(S'), S \middle| S_2 \tfrac{1}{2}(S_1)\tfrac{1}{2}, S\right) \times$$

$$\times (L_2, ll(L'), L \,|\, L_2\, l\,(L_1)\, l, L)\, (l^{n-2}(\alpha_2 T_2 S_2 L_2)\, l, T_1 S_1 L_1 |\} \, l^{n-1}\alpha_1 T_1 S_1 L_1) \times$$

$$\times (l^{n-1}(\alpha_1\, T_1\, S_1\, L_1)\, l, T\, S\, L \,|\} \, l^n\, \alpha\, T\, S\, L) = 0 \qquad (131)$$

for every value of T_2, S_2, L_2 and $T' + S' + L'$ even.

The interaction energy is given by

$$(l^n \alpha T S L\,|E|\, l^n \alpha' T S L) = \frac{n}{n-2} \sum_{\alpha_1 \alpha_1' T_1 S_1 L_1} (l^n \alpha T S L\, \{|\, l^{n-1}(\alpha_1 T_1 S_1 L_1)\, l,$$

$$T S L)\, (l^{n-1} \alpha_1\, T_1 S_1 L_1 \,|E|\, l^{n-1}\, \alpha_1'\, T_1 S_1 L_1)\, (l^{n-1}(\alpha_1 T_1 S_1 L_1)\, l, \qquad (132)$$

$$T S L| \} \, l^n \alpha' T S L)\,.$$

Although this procedure has been used successfully to calculate all atomic configurations d^n and the configuration f^3, it becomes extremely laborious for the higher configurations, and it is at this point that group theory comes to our aid in the following three ways:

a) The hitherto unspecified variable α will be replaced by a set of quantum numbers which is almost complete. The choice of these quantum numbers, suggested by group theory, will greatly simplify the calculations.

b) The c. f. p. will be calculated without the use of the cumbersome equations (131).

c) The summations in (132) will be simplified.

3. The classification of the states of l^n

The states of a single particle in a given shell are characterized by the set of quantum numbers $m_\tau\, m_s\, m_l$. There are $4(2l+1)$ independent states to which correspond the $4(2l+1)$ eigenfunctions $\Phi(m_\tau\, m_s\, m_l)$. If we have n particles in the same shell, the configuration has $\binom{4(2l+1)}{n}$ independent antisymmetrical states to which correspond the eigenfunctions $\psi(l^n \Gamma)$, where Γ is a set of quantum numbers which may assume $\binom{4(2l+1)}{n}$ different values.

If we consider the $\Phi(m_\tau\, m_s\, m_l)$ as the basis vectors of the $4(2l+1)$-dimensional space of the states of a single particle in a given shell, the $\psi(l^n \Gamma)$ will form a complete set of antisymmetrical tensors of rank n in this space. This means that a unitary transformation

$$\Phi'(m_\tau'\, m_s'\, m_l') = \sum_{m_\tau m_s m_l} \Phi(m_\tau\, m_s\, m_l)\, c(m_\tau\, m_s\, m_l;\, m_\tau'\, m_s'\, m_l') \qquad (133)$$

on the Φ's will induce in the ψ's the transformation

$$\psi'(l^n\, \Gamma') = \sum_\Gamma \psi(l^n\, \Gamma)\, c(\Gamma,\, \Gamma')\,. \qquad (134)$$

The $\psi(l^n \Gamma)$ are therefore the basis of a representation $\mathscr{A}_n$ of degree $\binom{4(2l+1)}{n}$ of the unitary group $U_{4(2l+1)}$, characterized by the partition

$$n = 1 + 1 + 1 + \cdots + 1 + 0 + 0 + \cdots + 0 \,.$$

In order to obtain a set of functions $\psi(l^n \, \Gamma)$ which will make the matrix of the perturbation energy as nearly diagonal as possible, we have to restrict the group $U_{4(2l+1)}$ to its largest subgroup under which the perturbation energy is invariant. If we assume that the interaction between particles is central and charge-independent, this group is the group of three independent rotations in the coordinate space, spin space, and isotopic spin space. If we proceeded in this way, which is traditional in the application of group theory to quantum mechanics, we should obtain the group theoretical definitions of only the six quantum numbers $T \, S \, L \, M_T \, M_S \, M_L$.

But since we want to obtain a more nearly complete set of quantum numbers, even though they may not be *good* quantum numbers, we shall rather carry out the transition from $U_{4(2l+1)}$ to $R_3 \times R_3 \times R_3$ by successive steps. We shall therefore impose successive restrictions on the group $U_{4(2l+1)}$ to obtain subspaces of the $\binom{4(2l+1)}{n}$-dimensional space of the representation $\mathscr{A}_n$ which will be invariant with respect to different subgroups. These subspaces are characterized by the highest weights of the representations to which they belong, and these highest weights will be our new quantum numbers.

We shall start by considering the subgroup of $U_{4(2l+1)}$ which consists of those transformations (133) which are of the form

$$c(m_\tau \, m_s \, m_l; \; m_\tau' \, m_s' \, m_l') = \gamma(m_\tau \, m_s; \; m_\tau' \, m_s') \, \bar{c}(m_l \, m_l') \,, \qquad (135)$$

with γ and $\bar{c}$ unitary; this subgroup is the direct product $U_4 \times U_{2l+1}$. If we denote by $\mathscr{H}_\Sigma$ the irreducible representations of U_{2l+1} and by $\mathscr{G}_{\Sigma'}$ those of U_4, the irreducible representations of $U_4 \times U_{2l+1}$ will be the Kronecker products

$$\mathscr{G}_{\Sigma'} \times \mathscr{H}_\Sigma \,. \qquad (136)$$

Every irreducible representation of $U_{4(2l+1)}$ is a reducible representation of $U_4 \times U_{2l+1}$ and breaks up into representations (136); the general law of decomposition is somewhat complicated, but for the particular case of the representation $\mathscr{A}_n$ it is very simple: only those representations (136) appear in the decomposition of $\mathscr{A}_n$ for which Σ' is Σ^*, the partition dual to Σ, and every representation of this kind occurs only once. Since the Young diagram which illustrates the partition Σ has not more than $2l + 1$ rows, the length of a row in the diagram of Σ^* cannot exceed $2l + 1$. Similarly, the length of a row of Σ cannot exceed four.

If the elements of the basis of $\mathscr{G}_{\Sigma^*}$ are characterized by a set of quantum numbers Θ and those of $\mathscr{H}_\Sigma$ by a set Δ, the elements of the basis of $\mathscr{G}_{\Sigma^*} \times \mathscr{H}_\Sigma$ will be characterized by the set $\Theta \, \Delta$, and the states of l^n by the set $\Sigma \, \Theta \, \Delta$.

As a second step we restrict U_{2l+1} to the orthogonal subgroup R_{2l+1} which leaves invariant the bilinear symmetric form

$$\sum_{m_l} (-)^{m_l} \, \Phi_1(m_l) \, \Phi_2(-m_l) \, ; \tag{137}$$

R_{2l+1} has R_3 as a subgroup because (137) is proportional to the eigenfunction of the S-state of l^2, which is left invariant by R_3. Let the irreducible representation of R_{2l+1} whose highest weight is W be $\mathscr{B}_W$. Then in R_{2l+1},

$$\mathscr{H}_\Sigma = \sum_W b_W \, \mathscr{B}_W \, . \tag{138}$$

The possibility that we may have $b_W > 1$ gives rise to a running index β to number the equivalent representations, but in practice β takes on only small values. (For the states d^n, $b_W \leqq 2$.)

The next step of the reduction is to restrict the orthogonal matrices $\bar{c}(m_l m_l')$ to the particular matrices belonging to the representation $\mathscr{D}_l$ of R_3. When this is done, every $\mathscr{B}_W$ becomes a representation of R_3 and will in general decompose as

$$\mathscr{B}_W = \sum_L c_L \, \mathscr{D}_L \, , \tag{139}$$

where L is the highest weight of $\mathscr{D}_L$. If $c_L > 1$, another running index γ distinguishes the various $\mathscr{D}_L$ belonging to a given L. We have thus arrived at the following scheme

$$\psi(l^n \, \Sigma \, \Theta \, \beta \, W \, \gamma \, L \, M_L) \, . \tag{140}$$

For the nuclear configuration d^n, c_L is never larger than three, but for higher values of l it is expected to be much larger. In the particular case of the configuration f^n we may avoid such large values for γ if we avail ourselves of the fortunate coincidence that when $l = 3$ there exists another group, contained in R_7 and containing the representation $\mathscr{D}_3$ of R_3 which is a realization of G_2 and may be used to introduce a new subclassification*.

In order to complete the scheme (140), we must now perform an analogous reduction for U_4. If we restrict it to its unimodular subgroup, we obtain the semi-simple group belonging to the vector diagram A_3, which we have seen to be the same as D_3. Therefore the unitary unimodular group in four dimensions is isomorphic with R_6. Applying to $\Sigma^* \equiv [\Lambda_1, \Lambda_2, \Lambda_3, \Lambda_4]$ the transformation (68), we obtain as highest weight of the representation of R_6

$$P = \frac{1}{2}(\Lambda_1 + \Lambda_2 - \Lambda_3 - \Lambda_4) \, , \quad P' = \frac{1}{2}(\Lambda_1 - \Lambda_2 + \Lambda_3 - \Lambda_4) \, ,$$

$$P'' = \frac{1}{2}(\Lambda_1 - \Lambda_2 - \Lambda_3 + \Lambda_4) \, . \tag{141}$$

This transformation, introduced by WIGNER, corresponds to considering instead of the unimodular unitary group in four dimensions the homomorphic group R_6 which is the group of rotations in the six-dimensional space of the spin and isotopic spin.

---

* See reference [23], section 4, subsection 3.

From R_6 we go on to the subgroup $R_3 \times R_3$ by restricting the transformations γ of (135) to those which are of the form

$$\gamma\,(m_\tau\,m_s;\,m_\tau'\,m_s') = \gamma_1(m_\tau;\,m_\tau')\,\gamma_2(m_s;\,m_s') \,. \tag{142}$$

In this subgroup the representations $\mathscr{G}_{\Sigma^*}$ may be decomposed

$$\mathscr{G}_{\Sigma^*} \equiv \mathscr{G}_{P\,P'\,P''} = \sum_{T\,S} a_{T\,S}\, \mathscr{D}_T \times \mathscr{D}_S \,.$$

In case $a_{T\,S} > 1$ we have to introduce a new running index α; this gives us finally a scheme for the wave functions of the entire shell

$$\psi\,(l^n\,\Sigma\,\alpha\,T\,S\,M_T\,M_S\,\beta\,W\,\gamma\,L\,M_L) \,. \tag{143}$$

Except for the presence of the running indices $\alpha\,\beta\,\gamma$, we have found a complete set of quantum numbers, and we have achieved the first of the three purposes set forth in § 1.

4. The factorization of the coefficients of fractional parentage

The wave functions on the right of (130) transform according to $\mathscr{A}_{n-1} \times \mathscr{A}_1$; that on the left transforms according to $\mathscr{A}_n$. Thus it is evident that the c. f. p. are a rectangular part of the matrices which perform the decomposition

$$\mathscr{A}_{n-1} \times \mathscr{A}_1 = \mathscr{A}_n + \cdots \,. \tag{144}$$

The calculation of these matrices is simplified by the following considerations:

We have seen that if a group g has a subgroup h, an irreducible representation $U_A(s)$ of g will in general be reducible in the subgroup h, consisting of elements t. Let us assume that the matrices $U_A(t)$ have been reduced

$$(\beta\,B\,b\,|U_A(t)|\,\beta'\,B'\,b') = (b\,|V_B(t)|\,b')\,\delta_{B\,B'}\,\delta_{\beta\,\beta'} \,, \tag{145}$$

where B specifies the different representations of h, b denotes their rows and columns, and β is a running index distinguishing equivalent irreducible representations. The Kronecker product $U_{A_1} \times U_{A_2}$ of two irreducible representations A_1 and A_2 of g can be completely reduced

$$U_{A_1} \times U_{A_2} = \sum_A c_A\,U_A \tag{146}$$

by a similarity transformation with a matrix

$$(A_1\,\beta_1\,B_1\,b_1;\,A_2\,\beta_2\,B_2\,b_2\,|\,A_1\,A_2\,\alpha\,A\,\beta\,B\,b) \tag{147}$$

where the parameter α is a running index which enumerates the A's whenever a c_A is greater than 1 in (146). We shall now state without proof[*] a corollary to Schur's lemma which will enable us to express this matrix in a simpler way.

The matrix elements of the transformation (147) are the products of the matrix elements of the transformation which reduces the Kronecker

---

[*] For the proof see reference [23], section 3.

product $U_{B_1}(t) \times U_{B_2}(t)$ in h and coefficients which are independent of the b's:

$$(A_1 \beta_1 \, B_1 b_1; \, A_2 \beta_2 \, B_2 b_2 \mid A_1 A_2 \, \alpha A \, \beta B \, b)$$
$$= (B_1 b_1 \, B_2 b_2 \mid B_1 B_2 \, B b) \, (A_1 \, \beta_1 B_1; \, A_2 \, \beta_2 B_2 \mid A_1 A_2 \, \alpha A \, \beta B) \, . \tag{148}$$

If we take the representation of $\mathscr{A}_n$ in the scheme (143) and apply this lemma to each subgroup of the chain which was constructed in the preceding section, we can bring the matrix which reduces the direct product (144) into the form

$$\begin{aligned}
&\Big(l^{n-1} (\Sigma_1 \alpha_1 \, T_1 S_1 \, M_{T_1} M_{S_1} \beta_1 W_1 \gamma_1 L_1 M_{L_1}); \frac{1}{2} \frac{1}{2} \, l m_\tau m_s m_l \mid \\
&\quad l^n \, \Sigma \alpha \, T S \, M_T M_S \, \beta W \, \gamma \, L \, M_L \Big) \\
&= \Big(T_1 \, M_{T_1} \frac{1}{2} \, m_\tau \Big| T_1 \frac{1}{2} \, T \, M_T \Big) \cdot \Big(S_1 \, M_{S_1} \frac{1}{2} \, m_s \Big| S_1 \frac{1}{2} \, S \, M_S \Big) \times \\
&\quad \times \Big(\Sigma_1^* \, \alpha_1 \, T_1 S_1; \, [1] \frac{1}{2} \frac{1}{2} \Big| \Sigma^* \, \alpha \, T \, S \Big) \times \\
&\quad \times (L_1 \, M_{L_1} \, l m_l \mid L_1 \, l \, L \, M_L) \cdot (W_1 \, \gamma_1 \, L_1; \, (1) \, l \mid W \, \gamma \, L) \times \\
&\quad \times (\Sigma_1 \, \beta_1 \, W_1; \, [1] \, (1) \mid \Sigma \, \beta \, W) \cdot (l^{n-1} \, \Sigma_1; \, l [1] \mid l^n \, \Sigma) \, ,
\end{aligned} \tag{149}$$

where the symbol (1) means $W = (1 \, 0 \ldots 0)$ and [1] means $\Sigma = [1 \, 0 \ldots 0]$. This expression as it stands is not exactly the c. f. p., since the wave functions on the right hand side of (130) contain already the M-dependent factors of (149); hence we have

$$\begin{aligned}
&(l^{n-1} (\Sigma_1 \, \alpha_1 \, T_1 \, S_1 \, \beta_1 \, W_1 \, \gamma_1 \, L_1) \, l, \, T \, S \, L \mid\} \, l^n \, \Sigma \, \alpha \, T \, S \, \beta \, W \, \gamma \, L) \\
&= \Big(\Sigma_1^* \, \alpha_1 \, T_1 S_1; \, [1] \frac{1}{2} \frac{1}{2} \Big| \Sigma^* \, \alpha \, T \, S \Big) (W_1 \gamma_1 L_1; \, (1) \, l \mid W \gamma L) \times \tag{150} \\
&\quad \times (\Sigma_1 \, \beta_1 \, W_1; \, [1] \, (1) \mid \Sigma \, \beta \, W) \, (l^{n-1} \, \Sigma_1; \, l [1] \mid l^n \, \Sigma) \, .
\end{aligned}$$

Thus the problem of calculating the c. f. p. is reduced to the separate calculation of the different factors which appear in this equation; but before we solve this last problem we have to develop a new mathematical tool.

5. The algebra of tensor operators

The algebra of vector operators and of their representation by matrices was developed by GÜTTINGER and PAULI* and is presented in standard form in Chapter III of CONDON and SHORTLEY (reference [15]). The possibility of extending it to tensors was indicated by ECKART and WIGNER**. We shall outline it here following reference [21], § 3, where the problem is treated by the standard methods of CONDON and SHORTLEY.

---

* Z. Physik **67**, 743 (1931).
** See references [17] and [18].

We define an irreducible tensor $\boldsymbol{T}^{(k)}$ of degree k to be a set of $2k + 1$ quantities $T_q^{(k)}$, $-k \leq q \leq k$, which under rotations in three-dimensional space transform like the $2k + 1$ spherical harmonics of degree k. If the operators J_x, J_y, J_z operate on these quantities, we have

$$J_z\, T_q^{(k)} = T_q^{(k)}\, (k\, q\, |J_z|\, k\, q) = q\, T_q^{(k)} \tag{151a}$$

$$(J_x \pm i J_y)\, T_q^{(k)} = T_{q\pm1}^{(k)}(k\, q \pm 1\, |J_x \pm i J_y|\, k\, q)$$
$$= \sqrt{(k \pm q + 1)\, (k \mp q)}\, T_{q\pm1}^{(k)}\,. \tag{151b}$$

If the $T_q^{(k)}$ are themselves operators, the left side of (151) must be replaced by commutators

$$[J_z,\, T_q^{(k)}] = T_q^{(k)}\, (k\, q\, |J_z|\, k\, q) = q\, T_q^{(k)} \tag{152a}$$

$$[J_x \pm i J_y,\, T_q^{(k)}] = T_{q\pm1}^{(k)}\, (k\, q \pm 1\, |J_x \pm i J_y|\, k\, q)$$
$$= \sqrt{(k \pm q + 1)\, (k \mp q)}\, T_{q\pm1}^{(k)}\,. \tag{152b}$$

As in vector algebra, it is possible to define many kinds of tensorial products. Guided by the example of the vector addition law in quantum mechanics we shall define the tensor product of order K by the equation

$$X_Q^{(K)} = \sum_{q_1 q_2} T_{q_1}^{(k_1)}\, U_{q_2}^{(k_2)}\, (k_1\, q_1\, k_2\, q_2 \,|\, k_1\, k_2\, K\, Q)\,, \tag{153}$$

and it is easy to verify that this satisfies (152). The unitarity of the Clebsch-Gordan coefficients permits us to solve this equation

$$T_{q_1}^{(k_1)}\, U_{q_2}^{(k_2)} = \sum_{KQ} X_Q^{(K)}\, (k_1 k_2\, K\, Q \,|\, k_1 q_1\, k_2 q_2)\,. \tag{154}$$

According to the definition (153) it would be logical to define a scalar product as $X_0^{(0)}$. However, it is traditional to define as scalar product the quantity

$$(\boldsymbol{T}^{(k)} \cdot \boldsymbol{U}^{(k)}) = \sum_q (-)^q\, T_q^{(k)} U_{-q}^{(k)} = (-)^k \sqrt{2k+1}\, X_0^{(0)}. \tag{155}$$

An example of this formula is the addition theorem for spherical harmoncis

$$P_k(\cos \omega_{12}) = \frac{4\pi}{2k+1} \sum_q (-)^q\, Y_{kq}(\theta_1\, \varphi_1)\, Y_{k-q}(\theta_2\, \varphi_2)\,, \tag{156}$$

where

$$\cos \omega_{12} = \cos\theta_1 \cos\theta_2 + \sin\theta_1 \sin\theta_2 \cos(\varphi_1 - \varphi_2)\,. \tag{156'}$$

If we represent the components of a tensor $T_q^{(k)}$ in the scheme $\alpha\, j\, m$ and write down (152) in the form of relations between matrices, (152a) tells us that the non-vanishing elements of $(\alpha\, j\, m\, |T_q^{(k)}|\, \alpha'\, j'\, m')$ satisfy the selection rule $m - m' = q$ and (152b) reduces to (124) if we replace $(\alpha\, j\, m\, |T_q^{(k)}|\, \alpha'\, j'\, m')$ by $(j'\, m'\, k\, q|\, j'\, k\, j\, m)$. Since (124) was sufficient to determine $(j'\, m'\, k\, q\, |\, j'\, k\, j\, m)$ apart from a normalization factor, we obtain

$$(\alpha\, j\, m\, |T_q^{(k)}|\, \alpha'\, j'\, m') = A\, (j'\, m'\, k\, q\, |\, j'\, k\, j\, m)\,, \tag{157}$$

with A independent of m and q.

In order to bring out the symmetries of the Clebsch-Gordan coefficients, it will be convenient to introduce the notation

$$(j_1\, m_1\, j_2\, m_2\, |\, j_1\, j_2\, j\, m) = (-)^{j+m} \sqrt{2j+1}\, V(j_1 j_2 j;\, m_1 m_2 - m) \tag{158}$$

where
$$V(a\,b\,c;\alpha\,\beta\,\gamma) = \delta_{\alpha+\beta+\gamma}\,\Delta\,(a\,b\,c)\cdot\sum_z (-)^{c-\gamma+z}\times$$

$$\times\frac{[(a+\alpha)!\,(a-\alpha)!\,(b+\beta)!\,(b-\beta)!\,(c+\gamma)!\,(c-\gamma!]^{\frac{1}{2}}}{z!\,(a+b-c-z)!\,(a-\alpha-z)!\,(b+\beta-z)!\,(c-b+\alpha+z)!}\times\quad (159)$$

$$\times\,(c-a-\beta+z)!\,,$$

and
$$\Delta\,(a\,b\,c) = \left[\frac{(a+b-c)!\,(a+c-b)!\,(b+c-a)!}{(a+b+c+1)!}\right]^{\frac{1}{2}}.\qquad (160)$$

The $V(a\,b\,c;\alpha\,\beta\,\gamma)$ thus defined have the symmetries

$$V(abc;\alpha\beta\gamma) = (-)^{a+b-c}\,V(bac;\beta\alpha\gamma) = (-)^{a+b+c}\,V(acb;\alpha\gamma\beta)$$
$$= (-)^{a+b+c}V(cba;\gamma\beta\alpha) = (-)^{2b}\,V(cab;\gamma\alpha\beta) = (-)^{2c}V(bca;\beta\gamma\alpha)\qquad (161\text{a})$$

and
$$V(abc;\alpha\beta\gamma) = (-)^{a+b+c}\,V(abc;-\alpha-\beta-\gamma)\,,\qquad (161\text{b})$$

and they vanish if a, b, c do not satisfy the triangle inequality, or if one of the numbers $a-|\alpha|$, $b-|\beta|$, $c-|\gamma|$ is negative. Further, they satisfy the orthogonality relations

$$\sum_{\alpha\beta} V(abc;\alpha\beta\gamma)\,V(abc';\alpha\beta\gamma') = \frac{1}{2c+1}\,\delta_{cc'}\,\delta_{\gamma\gamma'}\quad\text{or}\quad 0,\qquad (162)$$

$$\sum_{c\gamma}(2c+1)\,V(abc;\alpha\beta\gamma)\,V(abc;\alpha'\beta'\gamma) = \delta_{\alpha\alpha'}\,\delta_{\beta\beta'}\quad\text{or}\quad 0,\qquad (163)$$

the zeros occuring if any of the above conditions are violated by the parameters on which there is no summation.

In terms of the V's thus defined, we write (157) as

$$(\alpha j m\,|T_q^{(k)}|\,\alpha'j'm') = (-)^{j+m}(\alpha j\,\|T^{(k)}\|\,\alpha'j')\,V(jj'k;-mm'q)\quad (164)$$

This equation divides the physical properties of the tensor, which are described by $(\alpha j\,\|T^{(k)}\|\,\alpha'\,j')$ from its geometrical properties as described by the V's.

As an example of the utility of this separation we calculate the matrices of the scalar product (155) and find out by (162) that

$$(\alpha j m\,|(\boldsymbol{T}^{(k)}\cdot\boldsymbol{U}^{(k)})|\,\alpha'j'm')$$
$$= \frac{1}{2j+1}\,\delta_{jj'}\,\delta_{mm'}\sum_{\alpha''j''}(-)^{j-j''}(\alpha j\,\|T^{(k)}\|\,\alpha''j'')\,(\alpha''j''\,\|U^{(k)}\|\,\alpha'j')\,,\qquad (165)$$

which is, as required for a scalar, diagonal in j and m and independent of m.

In the practical applications, the most important scalar products are those in which the two tensors operate on different parts of a system. [Examples are $\boldsymbol{L}_1\cdot\boldsymbol{L}_2$, describing a coupling of the orbital angular momentum of two particles, $P_k\,(\cos\omega_{12})$ expressed by (156), or $\boldsymbol{L}\cdot\boldsymbol{S}$, in which space and spin functions belonging to the same system are coupled.] If $\boldsymbol{T}^{(k)}$ operates on part 1 of the system and $\boldsymbol{U}^{(k)}$ operates on part 2, then expressed in the scheme $\alpha_1\,\alpha_2 j_1 j_2 j\,m$, such a product is

$$(\alpha_1\,\alpha_2 j_1 j_2 j\,m\,|(\boldsymbol{T}^{(k)}\cdot\boldsymbol{U}^{(k)})|\,\alpha_1'\,\alpha_2'j_1'j_2'j'\,m')$$
$$= \sum_{m_1 m_1' m_2 m_2' q}(-)^q\,(j_1 j_2 j\,m\,|j_1\,m_1\,j_2\,m_2)\,(\alpha_1 j_1\,m_1|\,T_q^{(k)}|\,\alpha_1'j_1'\,m_1')\times\qquad (166)$$
$$\times\,(\alpha_2 j_2\,m_2\,|U_{-q}^{(k)}|\,\alpha_2'j_2'\,m_2')\,(j_1'\,m_1'j_2'\,m_2'\,|\,j_1'j_2'j'\,m')\,.$$

With (158) and (164), this involves sums over the products of four V's; it is found in general that

$$\sum_{\alpha\beta\gamma\delta\varphi} (-)^{f+\varphi} V(abc; \alpha\beta-\varepsilon) V(acf; -\alpha\gamma\varphi) V(bdf; -\beta\delta-\varphi) \times$$
$$\times V(cdg; \gamma\delta-\eta) = \frac{(-)^{e+g+f+d-b}}{2e+1} W(abcd; ef) \delta_{eg} \delta_{\varepsilon\eta}, \qquad (167)$$

where

$$W(abcd; ef) = \Delta(abe) \Delta(cde) \Delta(acf) \Delta(bdf) \sum_z (-)^z \times$$
$$\times \frac{(a+b+c+d+1-z)!}{(a+b-e-z)!\,(c+d-e-z)!\,(a+c-f-z)!\,(b+d-f-z)!} \times \qquad (168)$$
$$\times z!\,(e+f-a-d+z)!\,(e+f-b-c+z)!$$

Using (167) we obtain for (166) the expression

$$(\alpha_1 \alpha_2 j_1 j_2 j\, m\,|(\boldsymbol{T}^{(k)}\cdot\boldsymbol{U}^{(k)})|\,\alpha_1' \alpha_2' j_1' j_2' j'\, m')$$
$$= (-)^{j_1+j_2'-j} (\alpha_1 j_1 \|T^{(k)}\| \alpha_1' j_1') (\alpha_2 j_2 \|U^{(k)}\| \alpha_2' j_2') \times \qquad (169)$$
$$\times W(j_1 j_2 j_1' j_2'; j\, k) \delta_{jj'} \delta_{mm'}.$$

The geometrical interpretation of this formula is the following. If $\boldsymbol{T}^{(k)}$ is a 2^k-pole moment whose average (expectation value) in the direction of j_1 is $(\alpha_1 j_1 \|T^{(k)}\| \alpha_1 j_1)/\sqrt{2j_1+1}$ and similarly for $\boldsymbol{U}^{(k)}$ with j_2, then the diagonal elements of their scalar product are given in the limit of large j_1 and j_2 and small k by the product of these average values with $P_k(\widehat{j_1 j_2})$, where $\widehat{j_1 j_2}$ is the angle between j_1 and j_2; indeed, the asymptotic value of $(-)^{j_1+j_2-j} W(j_1 j_2 j_1 j_2; j\, k)$ in (169) is just equal to $P_k(\widehat{j_1 j_2})/\sqrt{(2j_1+1)(2j_2+1)}$.

Also, the W's have many symmetries,

$$W(abcd; ef) = W(badc; ef) = W(cdab; ef) = W(acbd; fe)$$
$$= (-)^{e+f-a-d} W(ebcf; ad) = (-)^{e+f-b-c} W(aefd; bc). \qquad (170)$$

The W's are useful also for expressing in the scheme $\alpha j_1 j_2 j\, m$ the components of a tensor which operates on part 1 or part 2. The matrix elements of $T_q^{(k)}$ are

$$(\alpha j_1 j_2 j\, m\, |T_q^{(k)}|\, \alpha' j_1' j_2' j'\, m') = \sum_{m_1 m_1' m_2} (j_1 j_2 j\, m\,|\, j_1 j_2 m_1 m_2) \times$$
$$\times (\alpha j_1 m_1\,|T_q^{(k)}|\, \alpha' j_1' m_1') (j_1' j_2 m_1' m_2\,|\, j_1' j_2 j'\, m');$$

using (164), (167) and the orthogonality relations of the V's, we get

$$(\alpha j_1 j_2 j\, \|T^{(k)}\| \alpha' j_1' j_2 j')$$
$$= (-)^{j_2+k-j_1'-j} \sqrt{(2j+1)(2j'+1)} \, (\alpha j_1 \|T^{(k)}\| \alpha' j_1') \, W(j_1 j j_1' j'; j_2 k). \qquad (171)$$

Analogously for $\boldsymbol{U}^{(k)}$

$$(\alpha j_1 j_2 \|U^{(k)}\| \alpha' j_1 j_2' j')$$
$$= (-)^{j_1+k-j_2-j'} \sqrt{(2j+1)(2j'+1)} \, (\alpha j_2 \|U^{(k)}\| \alpha' j_2') \, W(j_2 j j_2' j'; j_1 k). \qquad (172)$$

The geometrical interpretation of (171) and (172) is the same as that of (169).

A further use of the W's is to express the transformation connecting different schemes of parentage★

$$(j_1 j_2 (j_{12}) j_3, J \mid j_1, j_2 j_3 (j_{23}) J)$$
$$= \sqrt{(2j_{12} + 1)(2j_{23} + 1)} \ W(j_1 j_2 J j_3; j_{12} j_{23}) \ . \tag{173}$$

In general, every quantity which is invariant under rotations in three dimensions and therefore does not depend on the choice of axes or on m can be expressed in terms of the double-barred matrices and the W's.

6. Tensor operators and Lie groups

In part III we were not able to construct the matrices which decompose the Kronecker product of two representations because we did not even possess a complete scheme, except, of course, in the case of the group O_3. Now that we have a nearly complete scheme, the way is open to a further attempt. But the scheme we have achieved is not that of the weights, in which the H_i are diagonal, but it is a new scheme characterizing the physical problem, and one in which $T \ S \ L \ M_T \ M_S \ M_L$ are diagonal. Together with the diagonality of the H_i we have also lost the selection rules for the operators E_α; it is, therefore, convenient also to change to a basis of the infinitesimal operators of the group which fits the new scheme better. We shall see that such a basis can be given in terms of an apropriate set of tensor operators.

Let us consider the *unit* tensor operators defined by

$$(n l \| u^{(k)} \| n' \ l') = \delta_{n n'} \ \delta_{l l'} \tag{174}$$

which connect only states within the same shell. The matrices of these operators are, by (164)

$$(l m \mid u_q^{(k)} \mid l m') = (-)^{l+m} V(l l k; -m m' q) ; \tag{175}$$

for every value of k there are $2k + 1$ matrices of this kind with $2l + 1$ rows and columns; since V vanishes for $k > 2l$, this gives a total of $(2l + 1)^2$ matrices for each l.

It is easy to verify that the tensor product of two u's, as defined by (153), is given by a tensor X which satisfies

$$(l \| X^{(K)} \| l') = (-)^{k_1 + k_2 - K} \sqrt{2K + 1} \ W(k_1 l k_2 l; l K) ,$$

and hence

$$X_Q^{(K)} = (-)^{k_1 + k_2 - K} \sqrt{2K + 1} \ W(k_1 l k_2 l; l K) \ u_Q^{(K)} , \tag{176}$$

while the commutator of two u's is given by

$$[u_{q_1}^{(k_1)} \ u_{q_2}^{(k_2)}] = 2 \sum_{K Q}{}' (-)^{k_1 + k_2 - K} \sqrt{2K + 1} \ W(k_1 l k_2 l; l K) \times$$
$$\times (k_1 k_2 K Q \mid k_1 q_1 k_2 q_2) \ u_Q^{(K)} \tag{177}$$

where the prime on the summation indicates that, owing to the symmetries of the Clebsch-Gordan coefficients, the sum is to be taken only over

★ Ref. [22], Eq. (4).

values of K for which $k_1 + k_2 - K$ is odd. (177) is of the form (14) and hence defines the structure of a Lie group.

In virtue of the orthogonality relations (162), the $(2l + 1)^2$ matrices (175) are linearly independent. Since they are of degree $2l + 1$, they form a linearly complete set of matrices of this degree; it follows that the structure defined by (177) is that of the full linear group in $2l + 1$ dimensions and of its unitary subgroup U_{2l+1}.

For a system of n particles we can define a set of $u_i^{(k)}$ ($i = 1, 2, \ldots, n$), each operating on one particle, and we can construct the symmetrical tensors

$$U^{(k)} = \sum_{i=1}^{n} u_i^{(k)} \tag{178}$$

operating on the whole system. It is evident that the $U_q^{(k)}$ also satisfy the commutation relations (177). The matrices of the $U_q^{(k)}$ in the scheme (143) will therefore be the representations $\mathscr{H}_\Sigma$ of the infinitesimal operators of U_{2l+1}.

From (177) it is also seen that commutators of tensors of odd degree are linear combinations of again only such tensors; hence, the tensors $U^{(k)}$ of odd degree are the infinitesimal operators of a subgroup of the group U_{2l+1}. It is easy to see that this subgroup is the orthogonal subgroup R_{2l+1} which leaves invariant the bilinear form (137), the eigenfunction of the S-state of l^2; indeed the matrix elements $(l^2 L M | U_q^{(k)} | l^2 S O)$ vanish according to the triangular condition unless $k = L$, and vanish for odd L because the two states have different parity. The matrices of $U_q^{(k)}$ with odd k in the scheme (143) will therefore be the representations $\mathscr{B}_W$ of the infinitesimal operators of R_{2l+1}.

According to (164) the problem of the construction of the representations of R_{2l+1} and U_{2l+1} is reduced to the construction of the double-barred matrices of the $U^{(k)}$, and the problem of constructing the factors $(W_1 \gamma_1 L_1; (1) l | W \gamma L)$ and $(\Sigma_1 \beta_1 W_1; [1] (1) | \Sigma \beta W)$ of (150) is reduced to the construction of the similarity transformation which decomposes these matrices for odd and even k respectively.

7. Calculation of the coefficients of fractional parentage

In order to calculate the factors $(W_1 \gamma_1 L_1; (1) l | W \gamma L)$ we have to construct for every odd $k < 2l$ the matrices

$$(W_1 \gamma_1 L_1 l L \| U^{(k)} \| W_1 \gamma_1' L_1' l L') , \tag{179}$$

where $U^{(k)} = U_1^{(k)} + u^{(k)}$. This can be done by using equations (171) and (172) if we already know $(W_1 \gamma_1 L_1 \| U_1^{(k)} \| W_1 \gamma_1' L_1')$. The transformation matrix which decomposes (179) is $(W_1 \gamma_1 L_1; (1) l | W \gamma L)$.

If we are not interested in the matrices of $U^{(k)}$ per se but only in these transformation coefficients, it is sufficient to choose one particular odd value of $k > 1$, e. g., $k = 3$. $k = 1$ does not serve our purpose because $U^{(1)}$ is proportional to L and is therefore already diagonal in our scheme.

As an example we shall calculate the coefficients

$$((20) L_1; (1) d | W L) \tag{180}$$

for the configuration d^n. We first construct

$$(d^2 L \| u_1^{(3)} + u_2^{(3)} \| d^2 L') \tag{181}$$

for which, using (174), (171), (172) and Table 1, we obtain the matrix

L\L'	S	P	D	F	G
S	0	0	0	0	0
P	0	0	0	$\sqrt{\dfrac{6}{5}}$	0
D	0	0	$-\dfrac{8}{7}$	0	$\dfrac{3\sqrt{10}}{7}$
F	0	$\sqrt{\dfrac{6}{5}}$	0	$-\sqrt{\dfrac{3}{5}}$	0
G	0	0	$\dfrac{3\sqrt{10}}{7}$	0	$\dfrac{3\sqrt{11}}{7}$

$$\tag{182}$$

which decomposes by rearrangement of rows and columns into matrices of the form $(W L \| U^{(3)} \| W L')$

$$((00)\, S \| U^{(3)} \| (00)\, S) = |0|$$

$$((11)\, L \| U^{(3)} \| (11)\, L') = \begin{array}{c|cc} & P & F \\ \hline P & 0 & \sqrt{6/5} \\ F & \sqrt{6/5} & -\sqrt{3/5} \end{array} \tag{183}$$

$$((20)\, L \| U^{(3)} \| (20)\, L') = \begin{array}{c|cc} & D & \\ \hline D & -8/7 & 3\sqrt{11}/7 \\ G & 3\sqrt{10}/7 & 3\sqrt{10}/7 \end{array}$$

where the identification of the values of W to which these constituents belong has been made by the use of the *branching laws* as explained in detail in reference [26].

Now it is possible to obtain by the same method

$$((20)\, L_1\, dL \| U_1^{(3)} + u^{(3)} \| (20)\, L_1'\, dL') \tag{184}$$

Table 1. $W(2L\, 2L';\, 23)$

L\L'	S	P	D	F	G
S	0	0	0	$\dfrac{1}{\sqrt{35}}$	0
P	0	0	$\dfrac{\sqrt{2}}{5\sqrt{7}}$	$\dfrac{1}{\sqrt{70}}$	$\dfrac{-1}{\sqrt{210}}$
D	0	$\dfrac{\sqrt{2}}{5\sqrt{7}}$	$\dfrac{4}{35}$	$\dfrac{\sqrt{3}}{35\sqrt{2}}$	$\dfrac{-1}{7\sqrt{2}}$
F	$\dfrac{1}{\sqrt{35}}$	$\dfrac{1}{\sqrt{70}}$	$\dfrac{\sqrt{3}}{35\sqrt{2}}$	$\dfrac{-\sqrt{3}}{14\sqrt{5}}$	$\dfrac{-\sqrt{11}}{14\sqrt{5}}$
G	0	$\dfrac{-1}{\sqrt{210}}$	$\dfrac{-1}{7\sqrt{2}}$	$\dfrac{-\sqrt{11}}{14\sqrt{5}}$	$\dfrac{-\sqrt{11}}{42}$

but we shall see that in order to calculate (180) it is sufficient to know the last row of (184) which is

$$((20) \, G \, d \, I \, \| U_1^{(3)} + u^{(3)} \| \, (20) \, L_1' \, d \, L') \, . \tag{185}$$

Table 2

$L' \rightarrow$	F	G	H	I
$W(462\,L';23)$	$\dfrac{-1}{\sqrt{210}}$	$\dfrac{-\sqrt{7}}{9\sqrt{10}}$	0	0
$W(464\,L';23)$	$\dfrac{-1}{3\sqrt{2310}}$	$\dfrac{-\sqrt{7}}{33\sqrt{10}}$	$\dfrac{-7}{33\sqrt{15}}$	$\dfrac{-14\sqrt{2}}{33\sqrt{65}}$
$W(262\,L';43)$	$\dfrac{-1}{\sqrt{210}}$	$\dfrac{-\sqrt{7}}{3\sqrt{110}}$	$\dfrac{-1}{\sqrt{165}}$	$\dfrac{-\sqrt{2}}{\sqrt{715}}$

By the use of (183) and Table 2 we obtain (185) as the sum of

$$A = ((20) \, G \, d \, I \, \| U_1^{(3)} \| \, (20) \, L_1' \, d \, L')$$

and

$$B = ((20) \, G \, d \, I \, \| u^{(3)} \| \, (20) \, G \, d \, L') \, .$$

			D					G		
L' / L	S	P	D	F	G	D	F	G	H	I
A	0	0	0	$\dfrac{\sqrt{39}}{7}$	$\dfrac{\sqrt{13}}{7}$	0	$\dfrac{\sqrt{13}}{7\sqrt{30}}$	$\dfrac{3\sqrt{13}}{\sqrt{770}}$	$\sqrt{\dfrac{13}{15}}$	$2\sqrt{\dfrac{26}{55}}$
B	0	0	0	0	0	0	$-\sqrt{\dfrac{13}{30}}$	$\sqrt{\dfrac{91}{110}}$	$-\sqrt{\dfrac{13}{15}}$	$\sqrt{\dfrac{26}{55}}$
$A+B$	0	0	0	$\dfrac{\sqrt{39}}{7}$	$\dfrac{\sqrt{13}}{7}$	0	$-\dfrac{\sqrt{78}}{7\sqrt{5}}$	$\sqrt{\dfrac{130}{77}}$	0	$3\sqrt{\dfrac{26}{55}}$

(185')

It is possible to deduce from the branching laws that $\mathscr{B}_{(20)} \times \mathscr{B}_{(10)}$ decomposes into $\mathscr{B}_{(10)} + \mathscr{B}_{(21)} + \mathscr{B}_{(30)}$, and that to these three representations belong the states D, $P\,D\,F\,G\,H$, and $S\,F\,G\,I$ respectively. Hence, the transformation matrix which decomposes (184) by bringing it into the form

$$(W\,L\,\|U^{(3)}\|\,W\,L') \tag{186}$$

will have the structure

$$
\begin{array}{c}
(10) \quad D \\
\left\{
\begin{array}{c}
P \\
D \\
(21) \; F \\
G \\
H
\end{array}
\right. \\
\left\{
\begin{array}{c}
S \\
(30) \; F \\
G \\
I
\end{array}
\right.
\end{array}
\begin{pmatrix}
\;S\;\;P\;\;D\;\;F\;\;G\;\;D\;\;F\;\;G\;\;H\;\;I\; \\
* \quad\quad\quad\quad\quad * \\
\quad 1 \\
* \quad\quad\quad\quad\quad * \\
\quad\quad * \quad\quad\quad\quad * \\
\quad\quad\quad * \quad\quad\quad\quad * \\
\quad\quad\quad\quad\quad\quad\quad\quad 1 \\
1 \\
\quad\quad * \quad\quad\quad\quad * \\
\quad\quad\quad * \quad\quad\quad\quad * \\
\quad\quad\quad\quad\quad\quad\quad\quad\quad 1
\end{pmatrix} \tag{187}
$$

where the stars denote the non-vanishing matrix elements which have to be calculated. It follows from the form of (187) that the row $(30)\,I$ of (186) is obtained simply by multiplying the row (185), which has the elements (185′), with the columns of the transpose of (187). The selection rule which follows from the requirement that (186) is to be decomposed, in conjunction with all the available orthogonality and reciprocity relations, permits us to determine, apart from arbitrary phases, all elements of the matrix (187). They are contained in Table 3.

Table 3. $(W L \mid W_1 L_1; (1) d)$

W \\ L	W_1 / L_1	(00) S	(10) D	(11) P	(11) F	(20) D	(20) G
(00)	S	0	1	0		0	
(10)	D	1	0	$\sqrt{\frac{3}{10}}$	$\sqrt{\frac{7}{10}}$	$\sqrt{\frac{5}{14}}$	$\sqrt{\frac{9}{14}}$
(11)	P	0	1	$-\sqrt{\frac{8}{15}}$	$-\sqrt{\frac{7}{15}}$	0	
(11)	F		1	$-\sqrt{\frac{1}{5}}$	$\sqrt{\frac{4}{5}}$		
(20)	D	0	1	0		0	
(20)	G		1				
(21)	P			$-\sqrt{\frac{7}{15}}$	$\sqrt{\frac{8}{15}}$	1	0
(21)	D			$\sqrt{\frac{7}{10}}$	$-\sqrt{\frac{3}{10}}$	$\sqrt{\frac{9}{14}}$	$-\sqrt{\frac{5}{14}}$
(21)	F	0		$-\sqrt{\frac{4}{5}}$	$-\sqrt{\frac{1}{5}}$	$-\sqrt{\frac{2}{7}}$	$-\sqrt{\frac{5}{7}}$
(21)	G			0	-1	$-\sqrt{\frac{10}{21}}$	$\sqrt{\frac{11}{21}}$
(21)	H			0	1	0	1
(30)	S					1	0
(30)	F			0		$\sqrt{\frac{5}{7}}$	$-\sqrt{\frac{2}{7}}$
(30)	G					$\sqrt{\frac{11}{21}}$	$\sqrt{\frac{10}{21}}$
(30)	I					0	1

By reciprocity we mean the relation

$$(W\gamma L W_1\gamma_1 L_1; l) \sqrt{\frac{2L_1+1}{2L+1}\frac{g_W}{g_{W_1}}} (-)^{L-L_1+x}(W_1\gamma_1 L_1 \mid W\gamma L; l) \quad (188)$$

where g_W and g_{W_1} are the degrees of the representations and x is a phase which may be chosen arbitrarily for every pair W, W_1 but which is independent of the L's. This relation is proved in reference [23], equation (46); since its proof is based on the fact that the identity representation appears in the decomposition of $\mathscr{B}_W \times \mathscr{B}_W$, such a relation does not hold for the unitary group.

The calculation of

$$(\Sigma \beta W \mid \Sigma_1 \beta_1 W_1; [1] (1)) \tag{189}$$

can be carried out in an analogous fashion, using $U^{(2)}$ rather than $U^{(3)}$. The result for the configuration d^3 is given in Table 4. For the coefficients of the spin functions, corresponding to the passage from R_6 to $R_3 \times R_3$, the infinitesimal operators of the two groups R_3 are $\tau_\xi, \tau_\eta, \tau_\zeta$ and $\sigma_x, \sigma_y, \sigma_z$; the infinitesimal operators of R_6 are, in addition to these, $X_{\varrho r} = \tau_\varrho \sigma_r$, which form a tensor, or better a double vector with one foot in the isotopic-spin space and one in the spin space. The construction of the matrices

$$(\Sigma^* \alpha T S \| X \| \Sigma^* \alpha_1 T_1 S_1) \tag{190}$$

and the decomposition of the corresponding Kronecker products can be done in the same way as before (for $n = 3$, see Table 5).

Table 4. $(\Sigma W \mid \Sigma_1 W_1; [1] (1))$ for d^3

Σ	W	[11] (11)	[20] (00)	[20] (20)
[111]	(11)	1	0	
[210]	(10)	1	$\sqrt{\dfrac{8}{15}}$	$-\sqrt{\dfrac{7}{15}}$
	(21)	1	0	1
[300]	(10)		$\sqrt{\dfrac{7}{15}}$	$\sqrt{\dfrac{8}{15}}$
	(30)	0	0	1

Table 5. $\left(\Sigma^* T S \mid \Sigma_1^* T_1 S_1; [1] \dfrac{1}{2}\dfrac{1}{2}\right)$ for $n = 3$

$\Sigma^*(P P' P'')$ $(2T+1, 2S+1)$	$\Sigma_1^*(P_1 P_1' P_1'')$ $(2T_1+1, 2S_1+1)$	[20] (111) (11)	[20] (111) (33)	[11] (100) (13)	[11] (100) (31)
[300] $\left(\dfrac{3}{2}\dfrac{3}{2}\dfrac{3}{2}\right)$	(22)	$\dfrac{1}{\sqrt{2}}$	$\dfrac{1}{\sqrt{2}}$	0	
	(44)	0	1		
[210] $\left(\dfrac{3}{2}\dfrac{1}{2}\dfrac{1}{2}\right)$	(22)	$\dfrac{-1}{\sqrt{2}}$	$\dfrac{1}{\sqrt{2}}$	$\dfrac{-1}{\sqrt{2}}$	$\dfrac{-1}{\sqrt{2}}$
	(24)	0	1	1	0
	(42)	0	1	0	1
[111] $\left(\dfrac{1}{2}\dfrac{1}{2}\dfrac{-1}{2}\right)$	(22)	0		$\dfrac{-1}{\sqrt{2}}$	$\dfrac{1}{\sqrt{2}}$

For the construction of the coefficients

$$(l^n \Sigma \mid l^{n-1} \Sigma_1; l[1]) \tag{191}$$

the calculation is based not on the properties of Lie groups, but rather on those of the permutation groups, π_n. The result, which we give here without proof, is very simple

$$(l^n \Sigma \mid l^{n-1} \Sigma_1; l[1]) = \pm \sqrt{\frac{g_{\Sigma_1}}{g_\Sigma}} \tag{192}$$

where g_Σ is the degree of the representation of π_n which is characterized by the partition Σ. The sign depends on the choice of sign made for the other coefficients. For $n = 3$ they are given in Table 6.

Table 6. $(l^3 \Sigma \mid l^2 \Sigma_1; l[1])$

$\Sigma \backslash \Sigma_1$	[11]	[20]
[111]	1	0
[210]	$\dfrac{-1}{\sqrt{2}}$	$\dfrac{1}{\sqrt{2}}$
[300]	0	1

Part V. The calculation of the energy matrix

1. The interaction of two particles

Since the interaction matrix for n particles is calculated according to (132) in terms of that for $n - 1$ particles, we must start by calculating the interaction energy for two particles. Let us first assume for simplicity that there is an ordinary spin-independent interaction (Wigner interaction), given by $J(r_{12}) = J(\sqrt{r_1^2 + r_2^2 - 2r_1 r_2 \cos\omega_{12}})$ between the two particles. We can expand this in Legendre polynomials of $\cos\omega_{12}$

$$J(r_{12}) = \sum_k J_k(r_1, r_2) P_k(\cos\omega_{12}) \tag{193}$$

so that, by the addition theorem (156), they can be expressed in terms of scalar products of tensors

$$J(r_{12}) = \sum_k J_k(r_1, r_2) (C_1^{(k)} \cdot C_2^{(k)}) \tag{194}$$

where

$$C_{iq}^{(k)} = \sqrt{\frac{4\pi}{2k+1}} Y_{kq}(\theta_i \varphi_i) \tag{195}$$

is the qth component of $C_i^{(k)}$.

The matrix giving the interaction of two particles is in general

$$(n_1 l_1 n_2 l_2 L M \mid J(r_{12}) \mid n_1 l_1 n_2 l_2 L M)$$

$$= \sum_k (l_1 l_2 L M \mid (C_1^{(k)} \cdot C_2^{(k)}) \mid l_1 l_2 L M) F^k \tag{196}$$

$$= \sum_k (-)^{l_1 + l_2 - L} (l_1 \| C^{(k)} \| l_1) (l_2 \| C^{(k)} \| l_2) W(l_1 l_2 l_1 l_2; L K) F^k.$$

where the coefficients F^k are given by

$$F^k = \int\int J_k(r_1, r_2) R^2_{n_1 l_1}(r_1) R^2_{n_2 l_2}(r_2) \, dr_1 \, dr_2 , \tag{197}$$

which is called the *generalized Slater integral*.

It is interesting to note that the classical Slater integrals, which were defined for $J(r_{12}) = e^2/r_{12}$ are decreasing functions of k. But, if instead of a Coulomb interaction we have a short range interaction, F^k may no longer decrease with k. On the contrary, it is easy to see that for $J(r_{12}) = \delta(\overrightarrow{r_1} - \overrightarrow{r_2})$ which is the limiting case of short range interaction, one has

$$F^k = (2k + 1) F^0 . \tag{198}$$

For the particular case in which we are interested, of two particles in the same shell, (196) reduces to

$$(l^2 \, L \, M \, |J(r_{12})| \, l^2 \, L \, M) = \sum_k (-)^L (l \, \|C^{(k)}\| \, l)^2 \, W(l\,l\,l\,l; L \, k) \, F^k . \tag{199}$$

If, instead of a Wigner interaction we have some kind of exchange interaction, the sign of this expression has to be changed for some values of T, S, and L.

2. The group-theoretical classification of the interactions

The general formula for calculating the energy matrix for a system of n equivalent particles was given by (132), but since the α's stand for a set of many quantum numbers which may assume many different values, the summation of (132) is very long and has to be split up into a set of independent smaller summations. This is made possible by the factorization of the c. f. p. and by a similar factorization of the energy matrix which we shall discuss now.

We have seen in § 5 of part IV that there is a relation between the Clebsch-Gordan coefficients and the matrix elements of the components of the irreducible tensor operators. But the relation (157), which is a property of the group R_3, may be generalized to any other group if we adopt the general standpoint of ECKART and WIGNER.

If G is a group whose irreducible representations X have rows and columns characterized by χ, and $T(\Omega \, \omega)$ is an operator which has the same transformation properties with respect to the group as the element ω of the basis of the representation Ω of G, then, in analogy to (157), the matrix element $(X' \, \chi' \, |T(\Omega \, \omega)| \, X \, \chi)$ will be proportional to the matrix element $(X \Omega \, X' \, \chi' \, | \, X \chi \, \Omega \, \omega)$ of the transformation which decomposes the Kronecker product $X \times \Omega$. In the particular case that G is the group $U_{4(2l+1)}$ and X is $\mathscr{A}_n$ (cf. p. 63), we have

$$(l^n \, \Gamma' \, |T(\Omega \, \omega)| \, l^n \, \Gamma) = C (\mathscr{A}_n \, \Omega \, \mathscr{A}_n \, \Gamma' \, | \, \mathscr{A}_n \, \Gamma \, \Omega \, \omega) , \tag{200}$$

and, if we assume for $\mathscr{A}_n$ and Ω the scheme (143), this matrix element may be factorized according to (148).

Since a central and charge-independent interaction is a scalar with respect to the three-dimensional rotations in coordinate space, spin space, and isotopic-spin space, it follows that the energy matrix is diagonal with respect to $T\,S\,L$ and is independent of $M_T\,M_S\,M_L$, as is well known.

Unfortunately, the interaction is not an irreducible tensor operator with respect to the group $U_{4(2l+1)}$ and to its subgroups which were used to classify the states of l^n. We shall, therefore, as a first step decompose the interaction operator into a sum of interactions which are components of irreducible tensor operators, and then calculate the energy matrices of these particular interactions, using (150) and the factorization which follows from (200) and (148) to simplify the summations (132).

In general, the interaction operator will be a tensor of some kind which is reducible with respect to $U_{4(2l+1)}$, and, if for the time being we limit ourselves to a spin-independent interaction (WIGNER or MAJORANA), it will be a scalar with respect to U_4 and a tensor with respect to U_{2l+1}.

In order to identify the irreducible parts of this tensor, we start by considering an operator which operates on the space coordinates of a single particle in a given shell. Since it has to be a linear transformation in the $2l + 1$-dimensional space, it will be a tensor of the second rank with one covariant and one contravariant index. It was stated on p. 59 that the components of a (contravariant) vector are the basis of the representation $\mathscr{H}_{[10\,...\,0]}$; analogously, the components of a covariant vector are the basis of the representation $\mathscr{H}_{[00\,...\,0-1]}$; hence the components of a mixed tensor of rank two are the basis of the reducible representation

$$\mathscr{H}_{[10\,...\,0]} \times \mathscr{H}_{[00\,...\,0-1]}\,, \tag{201}$$

which decomposes into $\mathscr{H}_{[0\,...\,0]} + \mathscr{H}_{[10\,...\,0-1]}$. (This decomposition corresponds to separating the trace from the traceless part of the mixed tensor).

The interaction between two particles is expressed according to (196) as a sum of products of operators operating on the two particles, and will, therefore, belong to the basis of the reducible representation

$$(\mathscr{H}_{[0\,...\,0]} + \mathscr{H}_{[10\,...\,0-1]}) \times (\mathscr{H}_{[0\,...\,0]} + \mathscr{H}_{[10\,...\,0-1]}) \tag{202}$$

of U_{2l+1}.

If we decompose the representation (202) into its irreducible components and adopt a scheme in which W and L are diagonal, as we did in § 3 of part IV for the classification of the states, then, since the interaction is a scalar in the three-dimensional space, it will appear as a linear combination of the different basis elements which are classified as S-states in this scheme.

Since $\mathscr{H}_{[10\,...\,0]}$ and $\mathscr{H}_{[0\,...\,0-1]}$ are in R_3 the representation $\mathscr{D}_l$, (201) is the representation $\mathscr{D}_l \times \mathscr{D}_l$ which decomposes into $\sum_{L=0}^{2l} \mathscr{D}_L$, and it follows that in the basis of (202) there are $2l + 1$ independent invariants with respect to R_3 which have various tensorial characters in

U_{2l+1} and R_{2l+1}. It may be shown by the branching laws that two of them are invariants, also, with respect to U_{2l+1} and R_{2l+1}. One is still an invariant with respect to R_{2l+1}, but with respect to U_{2l+1} it belongs to the representation with highest weight $[20\ldots0-2]$. The other scalars are, in the scheme (143), of the following kinds: $[20\ldots0-2]$ (22) S, $[20\ldots0-2]$ (40) S, $[110\ldots0-1-1]$ (22) S, $[110\ldots0-1-1]$ (1111) S.

The decomposition of the interaction (196) into its irreducible parts may be made in a general way based on the fractional parentages of the different representations, as was done for f^n in reference [23], § 6, 1, but we shall consider here only the configurations d^n and follow a more empirical method.

Any kind of spin-independent interaction $E^{(\lambda)}$ will be represented in the d^2 configuration by a diagonal matrix

$$(d^2\,L\,M\,|E^{(\lambda)}|\,d^2\,L\,M) = f^{(\lambda)}(L) \tag{203}$$

and we have to calculate $f^{(\lambda)}(L)$ for the different irreducible parts into which the interaction (199) decomposes. The result is tabulated here:

Name	Tensorial Character			Σ: [11] (11) W: (11) L: P	[11] (11) F	[20] (00) S	[20] (20) D	[20] (20) G
	Σ	W	L					
$E^{(\alpha)}$	[00000]	(00)	S	1	1	1	1	1
$E^{(\beta)}$	[00000]	(00)	S	—1	—1	1	1	1
$E^{(\gamma)}$	[2000-2]	(00)	S	0	0	—14	1	1
$E^{(\varepsilon)}$	[2000-2]	(22)	S	0	0	0	—9	5
$E^{(\zeta)}$	[110-1-1]	(22)	S	—7	3	0	0	0

This table was obtained as follows: according to Schur's lemma $f^{(\alpha)}$ and $f^{(\beta)}$ have to be constant for states belonging to the same value of Σ; any linear combination of them has this property and the choice is determined only by considerations of simplicity. $f^{(\gamma)}$ must be constant for states belonging to the same value of W; moreover, according to (200) and in virtue of the orthogonality of the transformation matrices, $f^{(\gamma)}$ has to be orthogonal to both $f^{(\alpha)}$ and $f^{(\beta)}$ [if we consider every value of L with its $(2L+1)$-fold degeneracy]. $f^{(\varepsilon)}$ and $f^{(\zeta)}$ must be orthogonal to $f^{(\alpha)}$, $f^{(\beta)}$ and $f^{(\gamma)}$; and, in addition $f^{(\varepsilon)}$ has to vanish for $\Sigma = [11]$ since $\mathcal{H}_{[11]}$ does not appear in the decomposition of the Kronecker product $\mathcal{H}_{[11]} \times \mathcal{H}_{[2000-2]}$ and $f^{(\eta)}$ has to vanish for $\Sigma = [20]$ because $\mathcal{H}_{[20]}$ does not appear in the decomposition of the Kronecker product $\mathcal{H}_{[20]} \times \times \mathcal{H}_{[110-1-1]}$. (These selection rules are the analogs for U_{2l+1} of the triangular conditions for R_3.)

The perturbation energy $E(L)$ of the configuration d^2 for an ordinary (WIGNER) interaction may be obtained from (199). In order to avoid the appearance of fractional coefficients we introduce the standard normalization *

$$F_0 = F^0, \quad F_2 = F^2/49, \quad F_4 = F^4/441, \tag{204}$$

---

* Reference [15], p. 177.

and write for the energy[*]

$$E(S) = F_0 + 14F_2 + 126F_4$$
$$E(P) = F_0 + 7F_2 - 84F_4$$
$$E(D) = F_0 - 3F_2 + 36F_4 \qquad (205)$$
$$E(F) = F_0 - 8F_2 - 9F_4$$
$$E(G) = F_0 + 4F_2 + F_4 .$$

These results may be expressed in terms of the irreducible interactions by

$$V_W = E^{(\alpha)} F_0 + \left(-\frac{7}{12} E^{(\alpha)} + \frac{35}{12} E^{(\beta)} - \frac{5}{6} E^{(\gamma)}\right) (F_2 + 9F_4) + $$
$$+ \frac{1}{2} (E^{(\epsilon)} - 3E^{(\zeta)}) (F_2 - 5F_4) . \qquad (206)$$

The corresponding expression for the Majorana interaction is obtained by interchanging $E^{(\alpha)}$ and $E^{(\beta)}$ and changing the sign of $E^{(\zeta)}$:

$$V_M = E^{(\beta)} F_0 + \left(\frac{35}{12} E^{(\alpha)} - \frac{7}{12} E^{(\beta)} - \frac{5}{6} E^{(\gamma)}\right) (F_2 + 9F_4) + $$
$$+ \frac{1}{2} (E^{(\epsilon)} + 3E^{(\zeta)}) (F_2 - 5F_4) . \qquad (207)$$

3. The calculation of the energy matrices

When we go from d^2 to d^n, the summation over the different pairs of particles can be carried out very simply for the interactions $E^{(\alpha)}$ and $E^{(\beta)}$

$$\sum_{i<k} E_{ik}^{(\alpha)} = \frac{1}{2} n(n-1) \qquad (208a)$$

$$\sum_{i<k} E_{ik}^{(\beta)} = M \qquad (208b)$$

where M is the eigenvalue in the unperturbed state of the Majorana operator and can be expressed as a function of the partition $\Sigma^* = (\Lambda_1, \Lambda_2, \Lambda_3, \Lambda_4)$[**]

$$M = -\frac{1}{2} [\Lambda_1(\Lambda_1 - 1) + \Lambda_2(\Lambda_2 - 3) + \Lambda_3(\Lambda_3 - 5) + \Lambda_4(\Lambda_4 - 7)] . \qquad (209)$$

We can also obtain $\sum_{i<k} E_{ik}^{(\gamma)}$ in closed form by using the Casimir operator for the group R_5: it follows from (106) that for R_5 the eigenvalues of this operator are

$$g(W) = W_1(W_1 + 3) + W_2(W_2 + 1) , \qquad (210)$$

so that in particular,

$$g(00) = 0 , \quad g(10) = 4 , \quad g(11) = 6 , \quad g(20) = 10 . \qquad (210')$$

---

[*] Reference [15], p. 202.
[**] L. ROSENFELD, Nuclear Forces, Amsterdam, 1948, p. 211, Eq. (14).

We can therefore write

$$E^{(\gamma)} = \frac{3}{2}\left[g(W) - 2g(10)\right] - \frac{5}{2}E^{(\beta)} + \frac{1}{2}E^{(\alpha)} \tag{211}$$

and then find for d^n ★

$$
\begin{aligned}
\sum_{i<k} E^{(\gamma)}_{ik} &= \frac{3}{2}\left[g(W) - ng(10)\right] - \frac{5}{2}\sum_{i<k} E^{(\beta)}_{ik} + \frac{1}{2}\sum_{i<k} E^{(\alpha)}_{ik} \\
&= \frac{3}{2}g(W) - \frac{5}{2}M + \frac{1}{4}n(n - 25) .
\end{aligned}
\tag{212}
$$

In a similar way it is easy to see that

$$E^{(\epsilon)} + E^{(\zeta)} = L(L + 1) - \frac{3}{2}g(W), \tag{213}$$

so that also for nd-particles,

$$\sum_{i<k}(E^{(\epsilon)}_{ik} + E^{(\zeta)}_{ik}) = L(L + 1) - \frac{3}{2}g(W) . \tag{214}$$

The calculation of the energy matrices for the interactions $E^{(\epsilon)}$ and $E^{(\zeta)}$ separately has to be made by the use of (132); actually, owing to (214) it is sufficient to calculate

$$X = \frac{1}{2}\sum_{i<k}(E^{(\epsilon)}_{ik} - E^{(\zeta)}_{ik}) , \tag{215}$$

which will be of the form

$$
(d^n \, \Sigma \, \beta W \, \gamma L \,|X|\, d^n \, \Sigma \, \beta' W' \, \gamma' L) = \sum_{\varrho=1}^{r} (\Sigma \, \beta W |A_\varrho| \Sigma \, \beta' W') \times \\ \times (W | \Psi_\varrho(\gamma \gamma' L)| W') .
\tag{216}
$$

Although from (148) one might expect the factorization on the right hand side to be a complete one, this is not so because (148) did not represent the most general case. It contains the implicit assumption that in the decomposition of $U_{B_1} \times U_{B_2}$ the representation U_B appears only once, and this was actually the case when U_{B_2} was a representation which had as its basis the states of one particle in a given shell. However, now that U_{B_2} is the representation to which the interaction operator belongs, viz. $\mathscr{D}_{(22)}$, we need (148) in its most general form ★★, and this still involves a summation. The number of terms in this summation, r, equals the number of times that $\mathscr{B}_W$ appears in the reduction of $\mathscr{B}_{W'} \times \mathscr{B}_{(22)}$; and it follows from the branching laws that it can never exceed three.

Introducing (216) and (150) in (132), we obtain

$$
\begin{aligned}
(d^n\Sigma\beta\,W\gamma L|X|d^n\Sigma\beta'\,W'\gamma'L) &= \frac{n}{n-2}\sum_{\Sigma_1\beta_1\beta_1'W_1W_1'\gamma_1\gamma_1'L_1\varrho_1} (d^n\Sigma|\,d^{n-1}\Sigma_1;d\,[1]) \times \\
&\times (\Sigma\,\beta W|\Sigma_1\,\beta_1 W_1;\,[1]\,(1))\,(W\,\gamma L\,|\,W_1\,\gamma_1 L_1;\,(1)\,d) \times \\
&\times (\Sigma_1\beta_1 W_1|A_{\varrho_1}|\Sigma_1\beta_1'W_1')\,(W_1\,|\,\Psi_\varrho(\gamma_1\gamma_1'L_1)\,|\,W_1')\,(W_1'\gamma_1'L_1;\,(1)\,d| \\
&\quad |W'\,\gamma'L)\,(\Sigma_1\,\beta_1'W_1';\,[1]\,(1)\,|\,\Sigma\,\beta'W')\,(d^{n-1}\,\Sigma_1;d\,[1]\,|\,d^n\,\Sigma) .
\end{aligned}
\tag{217}
$$

★ The proof is the same as that of the well known formula

$$\sum_{i<k}(\mathbf{l}_i \cdot \mathbf{l}_k) = \frac{1}{2}\left[L(L + 1) - n\,l(l + 1)\right] .$$

★★ Reference [23], § 3.

We perform at first the summation

$$\sum_{\gamma_1 \gamma_1' L_1} (W \gamma L \mid W_1 \gamma_1 L_1; (1)\, d) \, (W_1 \mid \Psi_{\varrho_1}(\gamma_1 \gamma_1' L_1) \mid W_1') \times$$
$$\times (W_1' \gamma_1' L_1; (1)\, d \mid W' \gamma' L) \tag{218}$$

which, owing to the tensorial properties of the Ψ_ϱ will be a linear combination of the $(W \mid \Psi_\varrho(\gamma \gamma' L) \mid W')$ with coefficients that are independent of $\gamma \gamma'$ and L

$$\sum_{\gamma_1 \gamma_1' L_1} (W \gamma L \mid W_1 \gamma_1 L_1; (1)\, d)\, (W_1 \mid \Psi_{\varrho_1}(\gamma_1 \gamma_1' L_1) \mid W_1')\, (W_1' \gamma_1' L_1; (1)\, d \mid$$
$$\mid W' \gamma' L) = \sum_\varrho (W \mid x_\varrho (W_1\, W_1'\, \varrho_1) \mid W')\, (W \mid \Psi_\varrho(\gamma\, \gamma'\, L) \mid W') . \tag{218'}$$

When the $(W \mid \Psi_\varrho(\gamma \gamma' L) \mid W')$ are known, in order to obtain the coefficients of the linear combination, it suffices to perform the summation (218) for only a few values of $\gamma \gamma'$ and L.

Then we calculate

$$(\Sigma \beta W \mid y_\varrho(\Sigma_1) \mid \Sigma \beta' W') = \sum_{\beta_1 \beta_1' W_1 W_1' \varrho_1} (\Sigma \beta W \mid \Sigma_1 \beta_1 W_1; [1]\,(1))\, (\Sigma_1 \beta_1 W_1 \mid$$
$$\mid A_{\varrho_1} \mid \Sigma_1 \beta_1' W_1')\, (W \mid x_\varrho\, (W_1 W_1' \varrho_1) \mid W')\, (\Sigma_1 \beta_1' W_1'; [1]\,(1) \mid \Sigma \beta W) \tag{219}$$

and obtain finally

$$(\Sigma \beta W \mid A_\varrho \mid \Sigma \beta' W') = \frac{n}{n-2} \sum_{\Sigma_1} (d^n \Sigma \mid d^{n-1} \Sigma_1; [1]\,(1))^2 \times$$
$$\times (\Sigma \beta W \mid y_\varrho(\Sigma_1) \mid \Sigma \beta' W') . \tag{220}$$

In the particular case of a δ-interaction, which is the limit of forces with very short range, it follows from (198) and (204) that $F_2 - 5F_4$ vanishes and, therefore, the energies of the Wigner interaction may in this particular case be expressed in closed form

$$V = \frac{5}{14} F_0 [n(n+3) + 4M - g(W)] \tag{221}$$

by introducing (208a), (208b) and (212) into (206). Further, Wigner and Majorana interactions become equal for a δ-interaction.

Even if the interaction is not a δ-function, but is still of short range (compared with the dimensions of the nuclei), as is the case for nuclear interactions, the most important contributions to the energy come from $E^{(\alpha)}$, $E^{(\beta)}$ and $E^{(\gamma)}$, and the lowest levels are those with the smallest values of $g(W)$. These levels belong to $W = (00)$ for even nuclei and to $W = (10)$ for odd nuclei. Since $\mathscr{B}_{(00)} \times \mathscr{B}_{(22)} = \mathscr{B}_{(22)}$ and $\mathscr{B}_{(10)} \times \mathscr{B}_{(22)} = \mathscr{B}_{(21)} + \mathscr{B}_{(22)} + \mathscr{B}_{(32)}$ it follows that for $W = W' = (00)$ or $W = W' = (10)$, r vanishes in (216), i. e., for the levels belonging to these values of W the diagonal element of X vanishes. It must be remembered here that the W are not *good* quantum numbers; however, they are fairly good quantum numbers for short range forces, so that it is possible to calculate the lowest level of a configuration d^n without calculating the matrix of X.

4. Spin-dependent interactions

As in our previous discussion, we shall limit ourselves to the d^n shell in discussing spin-dependent interactions of the Bartlett and Heisenberg types, although the method is applicable to any nuclear shell. In addition to the five spin-independent irreducible interactions tabulated on p. 79, we have now five which depend on the spin and which may be obtained in the same manner:

Name	Tensorial Character Σ'	Σ	W	L	^{33}P	^{11}P	^{33}F	^{11}F	^{31}S	^{13}S	^{31}D	^{13}D	^{31}G	^{13}G
$E^{(\eta)}$	[11-1-1]	[00000]	(00)	$''S$	0	0	0	0	1	—1	1	—1	1	—1
$E^{(\theta)}$	[200-2]	[00000]	(00)	$''S$	1	—9	1	—9	0	0	0	0	0	0
$E^{(\gamma')}$	[11-1-1]	[2000-2]	(00)	$''S$	0	0	0	0	—14	14	1	—1	1	—1
$E^{(\varepsilon')}$	[11-1-1]	[2000-2]	(22)	$''S$	0	0	0	0	0	0	—9	9	5	—5
$E^{(\zeta')}$	[200-2]	[110-1-1]	(22)	$''S$	—7	63	3	—27	0	0	0	0	0	0

(Interaction / States of d^2)

In this table Σ' characterizes the representation $\mathscr{G}_{\Sigma'}$ of U_4 and Σ characterizes the representation $\mathscr{H}_\Sigma$ of U_5 to which the interactions belong.

The Bartlett and Heisenberg interactions must now be expressed in terms of these interactions, and it is easy to see that

$$V_B + V_H = 2E^{(\eta)}\left[F_0 + \frac{7}{3}(F_2 + 9F_4)\right] - \frac{5}{3}E^{(\gamma')}(F_2 + 9F_4) + \\ + E^{(\varepsilon')}(F_2 - 5F_4)\,, \tag{222}$$

$$V_B - V_H = \frac{2}{5}(E^{(\theta)} + 2E^{(\alpha)} - 2E^{(\beta)})\left[F_0 - \frac{7}{2}(F_2 + 9F_4)\right] - \\ - \frac{3}{5}(E^{(\zeta')} + 4E^{(\zeta)})(F_2 - 5F_4)\,. \tag{223}$$

It is also easy to show that

$$2\sum_{i<k}E^{(\eta)}_{ik} = S(S+1) - T(T+1) \tag{224}$$

and

$$\frac{2}{5}\sum_{i<k}(E^{(\theta)}_{ik} + 2E^{(\alpha)}_{ik} - 2E^{(\beta)}_{ik}) = S(S+1) + \\ + T(T+1) + \frac{1}{2}n(n-4)\,. \tag{225}$$

The calculation of the energy matrices for the interactions $E^{(\gamma')}$, $E^{(\varepsilon')}$, and $E^{(\zeta')}$ has to be made by the methods used in the preceeding section for the interaction X.

References

To part I:

[1] LIE, S., and G. SCHEFFERS: Vorlesungen über kontinuierliche Gruppen. Leipzig 1893
[2] CARTAN, E.: Sur la structure des groupes de transformations finis et continue. Thèse, Paris 1894 — II. edition 1933

[3] Bianchi, L.: Lezioni sulla teoria dei gruppi continui finiti di trasformazioni. Pisa 1903 — II. edizione 1918
[4] Eisenhart, L. P.: Continuous groups of transformations. Princeton 1933, Dover publications, 1961
[5] Weyl, H.: The structure and representation of continuous groups I and II. Princeton 1934 and 1934/35 (Mimeographed notes)

To part II:
[6] Weyl, H.: Math. Z. 23, 271 (1925); 24, 328 and 377 (1926)
[7] van der Waerden, B. L.: Math. Z. 37, 446 (1933)
[8] Coxeter, H. S. M.: In Weyl [5], p. 186
[9] — Regular Polytopes. London: Methuen 1948

To part III:
[10] Cartan, E.: Bull. Soc. Math. France 41, 53 (1913)
[11] Casimir, H.: Proc. Kon. Acad. Amst. 34, 844 (1931)
[12] — and B. L. van der Waerden: Math. Ann. 111, 1 (1935)
[13] Weyl, H.: Classical Groups. Princeton 1939, second edition 1946
[14] Racah, G.: Rend. Lincei 8, 108 (1950)

To part IV:
[15] Condon, E. U., and G. H. Shortley: The Theory of Atomic Spectra. Cambridge 1935, reprinted with corrections 1959
[16] Weyl, H.: Gruppentheorie und Quantenmechanik. Second edition, Leipzig 1931. Dover publications (Translation into English)
[17] Eckart, C.: Rev. Mod. Phys. 2, 305 (1930)
[18] Wigner, E. P.: Gruppentheorie und ihre Anwendung auf die Quantenmechanik der Atomspektren. Braunschweig 1931. An expanded and improved edition in English was published by the Academic Press, New York and London 1959
[19] van der Waerden, B. L.: Die gruppentheoretische Methode in der Quantenmechanik. Springer, Berlin 1932
[20] Racah, G.: Phys. Rev. 51, 186 (1942)
[21] — Phys. Rev. 62, 438 (1942)
[22] — Phys. Rev. 63, 367 (1943)
[23] — Phys. Rev. 76, 1352 (1949)
[24] Wigner, E. P., and E. Feenberg: Reports on Progress in Physics VIII, 274 (1941). This review article contains references to all earlier theoretical papers on nuclear spectroscopy
[25] Racah, G.: Rev. Mod. Phys. 21, 494 (1949)
[26] Jahn, H. A.: Proc. Roy. Soc. A 201, 516 (1950)

To part V:
[27] Racah, G.: Helv. Phys. Acta 23, 229 (1950)

Prof. Giulio Racah
The Hebrew University
Jerusalem

Continuous Groups in Quantum Mechanics[*,**]

W. Pauli†

Contents

I. Continuous groups and their associated Lie algebra

An abstract group G is a set of elements which possess a *law of composition* such that from any two elements a, b of G taken in a given order arises an element c. The following conditions must be satisfied:

1) Associative law:

$$c(ba) = (cb)a .$$

2) Unit element: there exists in G an element I such that for any a of G

$$Ia = aI = a .$$

---

† W. Pauli died on 15 December 1958. The report was originally published as CERN-report 56—31. We are very obliged to Mrs. Pauli and CERN for the permission of reprinting (ed.).

* The present notes are based on a series of lectures given by Prof. W. Pauli, of the E. T. H., Zürich, to the members of the CERN Theoretical Study Division at Copenhagen in September 1955.

** The author is indebted to R. Jost and A. R. Edmonds for help in the preparation of these notes.

3) Every element a of G possesses an inverse a^{-1}, the composition of which with a gives the unit element:

$$aa^{-1} = a^{-1}a = I .$$

The continuous groups which occur in physics appear, however, not with such abstract elements but as groups of transformations of variables. For these the associative law 1) and the existence of the unity element 2) are automatically fulfilled and 3) means the one-one valuedness of the transformation. The continuous transformation groups in question can be divided into two classes:

1) The finite continuous groups, the elements of which are determined by a finite number of parameters. Examples: rotation group, Lorentz group.

2) The infinite continuous groups, in the elements of which occur arbitrary functions. Examples: gauge group, coordinate transformations in general relativity, canonical transformations in classical mechanics and in quantum mechanics.

These lectures will deal with the first class only; I wish, however, to emphasize that according to my opinion the second class will presumably turn out to be of greater importance for physics in future.

In order to treat the continuous groups of the first class we use the infinitesimal methods of S. Lie [1, 2]. For the groups of physics the conditions of differentiability are indeed always fulfilled for suitably chosen parameters. Many interesting questions can be treated already in the frame of the infinitesimal elements which define the so-called *Lie-ring*.

To define it we use as a starting point the fact that the group elements which deviate not too largely from the unit element I, can be parametrized in the following way: to every r-tuple of numbers $s^1, \ldots, s^r$ in an environment of $s^\alpha = 0$, $(\alpha = 1, \ldots, r)$ corresponds in a one-one valued way a group element which we shall simply denote by $(s^1, \ldots, s^r)$ or (s), with $I \sim (0)$. The number r is uniquely determined and is called the dimension of the group. To the product of two elements (s) and (t) there corresponds again a point u in the considered environment.

One has

$$u^\alpha = \varphi^\alpha(s^1, \ldots, s^r, t^1, \ldots, t^r) \tag{1}$$

where the functions φ^α can be assumed here to be differentiable an arbitrary finite number of times with respect to the arguments $s^1, \ldots, s^r$, $t^1, \ldots, t^r$. From $I \sim (0)$ it follows

$$\varphi^\alpha(0, t) = t^\alpha ; \quad \varphi^\alpha(s, 0) = s^\alpha . \tag{2}$$

The development of $\varphi^\alpha(s, t)$ is, therefore (compare Chap. 9 of Pontrjagin [3]),

$$\varphi^\alpha(s, t) = s^\alpha + t^\alpha + a^\alpha_{\beta\gamma} s^\beta t^\gamma + g^\alpha_{\beta\gamma\delta} s^\beta s^\gamma t^\delta + h^\alpha_{\beta\gamma\delta} s^\beta t^\gamma t^\delta + \cdots . \tag{3}$$

The commutator $q = (s)\,(t)\,(s^{-1})\,(t^{-1})$ of the two elements s and t is of fundamental importance. The development

$$q^\alpha(s, t) = c^\alpha_{\beta\gamma} s^\beta t^\gamma + \cdots \tag{4}$$

gives rise to the occurence of the structure constants $c_{\beta\gamma}^{\alpha}$ of the group, connected with the coefficients $a_{\beta\gamma}^{\alpha}$ in (3) by

$$c_{\beta\gamma}^{\alpha} = a_{\beta\gamma}^{\alpha} - a_{\gamma\beta}^{\alpha} .$$

Hence

$$c_{\beta\gamma}^{\alpha} = - c_{\gamma\beta}^{\alpha} . \tag{5}$$

One uses here that in the reciprocal $(s)^{-1} = (\tilde{s})$ the following development holds

$$\tilde{s}^{\alpha} = - s^{\alpha} + a_{\beta\gamma}^{\alpha} s^{\beta} s^{\gamma} + \cdots .$$

A further condition for the $c_{\beta\gamma}^{\alpha}$ follows from the associative law which is for our parametrisation formulated by

$$\varphi(\varphi(s, t), u) = \varphi(s, \varphi(t, u)) .$$

Inserting the development (3) for φ one sees that the equation is identically fulfilled in first and second order; in third order, however, one finds for the coefficient of $s^{\beta} t^{\gamma} u^{\delta}$ the condition

$$a_{\sigma\delta}^{\alpha} a_{\beta\gamma}^{\sigma} - a_{\beta\sigma}^{\alpha} a_{\gamma\delta}^{\sigma} = h_{\beta\gamma\delta}^{\alpha} + h_{\beta\delta\gamma}^{\alpha} - g_{\beta\gamma\delta}^{\alpha} - g_{\gamma\beta\delta}^{\alpha} . \tag{6}$$

Antisymmetrisation with respect to the indices β, γ, δ makes the right side vanish, while the left side can be expressed with the help of (4) by the $c_{\beta\gamma}^{\alpha}$ and gives rise to the condition

$$c_{\beta\sigma}^{\alpha} c_{\gamma\delta}^{\sigma} + c_{\gamma\sigma}^{\alpha} c_{\delta\beta}^{\sigma} + c_{\delta\sigma}^{\alpha} c_{\beta\gamma}^{\sigma} = 0 . \tag{7}$$

Besides this no further condition follows.

We can now proceed to the definition of the Lie-ring of the infinitesimal group elements.

If $(s(\tau))$ is a one-dimensional submanifold (curve) of the group, containing the unit element $s^{\alpha} = 0$ for $\tau = 0$, one considers the tangent of the curve for $\tau = 0$, supposed as existent and defined by

$$\xi^{\alpha} = \left(\frac{ds^{\alpha}}{d\tau}\right)_{\tau=0} .$$

The r-dimensional vector space of these tangent vectors ξ^{α} is the Lie-ring.

From the two curves $(s(\tau))$ and $(t(\tau))$ a third curve with the parameter

$$\tau' \equiv \tau^2$$

can be constructed by

$$(u(\tau')) = (s(\tau)) (t(\tau)) (s(\tau))^{-1} (t(\tau))^{-1} .$$

With $\xi^{\alpha} = \left(\frac{ds^{\alpha}}{d\tau}\right)_{\tau=0}$ and $\eta^{\alpha} = \left(\frac{dt^{\alpha}}{d\tau}\right)_{\tau=0}$ one has according to (4) the development

$$u^{\alpha}(\tau') = c_{\beta\gamma}^{\alpha} \xi^{\beta} \eta^{\gamma} \tau' + \cdots .$$

The tangent vector of this third curve is therefore given by

$$\zeta^{\alpha} \equiv [\xi, \eta]^{\alpha} = c_{\beta\gamma}^{\alpha} \xi^{\beta} \eta^{\gamma} . \tag{8}$$

With the help of r linearly independent basis vectors $e_{\alpha}(\alpha = 1, \ldots, r)$, one can also write a vector ξ with the components ξ^{α} as

$$\xi = \xi^{\alpha} e_{\alpha} .$$

Defining then

$$[e_\beta, e_\gamma] = c_{\beta\gamma}^\alpha\, e_\alpha \tag{8a}$$

and considering ξ^α, η^α as ordinary commuting c-numbers, one obtains agreement with (8) for

$$\xi = \xi^\alpha\, e_\alpha;\quad \eta = \eta^\alpha\, e_\alpha;\quad \zeta \equiv [\xi, \eta] = \zeta^\alpha\, e_\alpha\,.$$

The position of the indices in (8) or (8a) shows the transformation law by a transition from the basis e_α to a new basis e_α'. The existence of the bracket-operation is characteristic for a Lie-ring. It has the properties

$$\left.\begin{array}{l} [\xi, c_1\, \eta_1 + c_2\, \eta_2] = c_1\, [\xi, \eta_1] + c_2\, [\xi, \eta_2] \\[4pt] [\xi, \eta] + [\eta, \xi] = 0 \\[4pt] [[\xi, \eta]\, \zeta] + [[\eta, \zeta]\, \xi] + [[\zeta, \xi]\, \eta] = 0\,. \end{array}\right\} \tag{9}$$

The last property is the Jacobi-identity which follows from the associative law and its consequence (7) for the $c_{\beta\gamma}^\alpha$.

The mathematicians have proved that Lie-groups with the same structure constants $c_{\beta\gamma}^\alpha$ are isomorphic in the small (i. e. in a sufficiently small environment of the unit element) and that to every Lie-ring [the product $[\xi, \eta]$ of which has the properties (9)] there exists a Lie-group. (See for instance Pontrjagin [3].)

II. Representations of Lie-rings and Lie-groups

Suppose that to each element ξ^α of the Lie-ring there corresponds a linear operator $A(\xi)$ (linear transformation of a vector space, the dimension of which is in general different from r) with the properties

$$\left.\begin{array}{l} A(\xi + \eta) = A(\xi) + A(\eta) \\[4pt] A(c, \xi) = c \cdot A(\xi) \\[4pt] A([\xi, \eta]) = A(\xi) A(\eta) - A(\eta) A(\xi) \equiv [A(\xi), A(\eta)] \end{array}\right\} \tag{10}$$

the last of which fulfils the Jacobi-identity identically, as the multiplication of linear operators is associative. Such a correspondence of operators (linear transformations) with the elements of a Lie-ring is called a *representation* of the Lie-ring.

In all cases considered here a representation of the Lie-ring determines uniquely also a representation of the Lie-group, namely a one-to-one correspondence of linear operators $G(s)$ to the group-elements (s)

$$(s) \rightleftarrows G(s)$$

with the properties

$$(s)\,(t) \to G(s)\,G(t)$$

and

$$(s)^{-1} \to G^{-1}(s)$$

from which also follows $G(0) = I$.

We do not give here a general proof for the possibility of this extension, but the reverse passage from a representation of the Lie-group to a representation of the Lie-ring is trivial: if $(s(\tau))$ is a "smooth" curve, then $G(s(\tau))$ associates with it a curve $G(\tau)$ in the space of the operators considered. If now

$$\xi^\alpha = \left(\frac{ds^\alpha}{d\tau}\right)_{\tau=0}$$

then the operator

$$A(\xi) = \left(\frac{dG}{d\tau}\right)_{\tau=0}$$

can be made by definition to correspond to ξ^α. It can easily be seen that this correspondence fulfils all the conditions (10).

In the part of the lecture contained in these notes we are dealing with representations of Lie-rings rather than Lie-groups. The Lie-ring always permits a particular representation by r-dimensional linear transformations called the *adjoint representation*. One obtains it by interpreting $[\alpha, \xi]$ as a linear transformation of the vector ξ where α, β, . . ., denote elements of the Lie-ring, too

$$[\alpha, \xi] = A(\alpha)\,\xi. \tag{11}$$

This is a representation because

$$A(\beta)\,A(\alpha)\,\xi = [\beta\,[\alpha, \xi]]$$
$$A(\alpha)\,A(\beta)\,\xi = [\alpha\,[\beta, \xi]]$$

hence, in view of the Jacobi-identity,

$$[A(\alpha),\,A(\beta)]\,\xi = [\alpha\,[\beta, \xi]] - [\beta\,[\alpha, \xi]] = [[\alpha, \beta]\,\xi] = A([\alpha, \beta])\,\xi.$$

The adjoint representation is not necessarily faithful, which means that it is possible that $A(\alpha) = 0$ for $\alpha \neq 0$. This is so for those α's, for which $[\alpha, \xi] = 0$. These particular α's are again a Lie-ring to which the "*centre*" of the Lie-group corresponds.

Written in matrices the adjoint representation is given by the correspondence

$$\xi^k \rightarrow c^\alpha_{k\beta}$$

where α, β are row and column-indices of the matrix for each given k.

We mention here that to every representation $A(\xi)$ there exists a *contravariant* representation given by

$$A'(\xi) = -A^T(\xi)$$

where T is the transposed matrix obtained by permutation of rows and columns (for operators: the adjoint operator). For the Lie-group this corresponds to the substitution $(G^{-1})^T(s)$ for $G(s)$.

We use the following general *definitions* of group theory. A group is *Abelian* if all its elements commute. Clearly for an Abelian group all commutators and hence all structure constants vanish.

A *subgroup* of a group G is a subset of elements of G, which satisfy the group postulates.

An *invariant subgroup* N of a group G is a subgroup which contains all the conjugates of its elements: i. e. if n is any element of N and s any element of G, then $s n s^{-1}$ is in N.

A group is *simple* if it contains no invariant subgroup besides the unit element.

A group is *semisimple* if it has no Abelian invariant subgroup besides the unit element. In this case the adjoint representation is faithful.

We also say that the Lie-ring associated with a semisimple subgroup is *semisimple*. CARTAN has given a sufficient and necessary condition for the group or Lie-ring to be semisimple, which uses only properties of the Lie-ring: construct from the structure constants $c_{\beta\gamma}^{\alpha}$ the symmetric tensor

$$g_{ik} = g_{ki} = c_{i\sigma}^{\varrho} c_{k\varrho}^{\sigma} . \tag{12}$$

The condition in question is then, that the determinant of the g_{ik} is different from zero:

$$\det |g_{ik}| \neq 0 . \tag{13}$$

Summarizing we give the following key of translation of concepts related to Lie-groups to concepts related to Lie-rings:

subgroup $\leftrightarrow$ subring: $[\alpha, \beta]$ is in the subring together with α and β;

invariant subgroup $\leftrightarrow$ ideal J: if α in J, ξ in Lie-ring R, then $[\xi, \alpha]$ in J;

Abelian group $\leftrightarrow [\xi, \eta] = 0$;

semisimple group $\leftrightarrow \det |g_{ik}| \neq 0$.

For semisimple groups there exists a theorem of Casimir, which we shall use in the following. To formulate it, one considers the normalized subdeterminants g^{ik} of g_{ik}, defined by

$$g^{i\varrho} g_{\varrho k} = \delta_k^i \tag{14}$$

which exist if and only if $\det |g_{ik}| \neq 0$. Then one considers the quadratic form

$$F = g^{\varrho\sigma} e_\varrho e_\sigma \tag{15}$$

where the e_ϱ are again r linearly independent elements of the Lie-ring. The theorem of Casimir states then: F commutes with all elements of the Lie-ring or

$$[F, e_\tau] \equiv g^{\varrho\sigma} (e_\varrho [e_\sigma, e_\tau] + [e_\varrho, e_\tau] e_\sigma) = 0 . \tag{16}$$

Before we indicate the proof, we remark first that the ordinary product $e_\varrho e_\sigma$ is not defined inside the Lie-ring. It is only defined either for representations of the Lie-ring by linear operators (or matrices) or in some amplified algebra. We adopt here the first point of view which also justifies the definition of $[F, e_\tau]$, used in (16).

To prove Casimir's theorem, one shows first, using the condition (7) for the structure constants and the definition (12) of the g_{ik}, that

$$c_{ijk} \equiv g_{i\varrho} c_{jk}^{\varrho} \tag{17}$$

is antisymmetric in i and j,

$$c_{ijk} = -c_{jik} . \tag{18}$$

Raising the indices i and j with help of the g^{ik} by

$$c^{ij}_{,k} = g^{\sigma i}\, g^{\tau j}\, c_{\sigma\tau k} = g^{\tau j}\, c^i_{\tau k} \tag{19}$$

it follows then that $c^{ij}_{,k}$, too is antisymmetric in i and j,

$$c^{ij}_{,k} = -c^{ji}_{,k}\,. \tag{20}$$

From the expression for $[F, e_\tau]$ given in (16) it follows, however, by (8a) and (19)

$$[F, e_\tau] = c^{\varrho\sigma}_{,\tau}(e_\varrho\, e_\sigma + e_\sigma\, e_\varrho)$$

which vanishes in view of (20), q. e. d.

The theorem of complete reducibility of every finite reducible representation of semisimple Lie-rings (Lie-groups)

We wish to recall, firstly, the general concepts concerning representations which also apply to Lie-rings and Lie-groups (cf. [4, 5, 6, 7]).

Two representations $r \to A_1(r)$, $r \to A_2(r)$ are called equivalent if for every element r of the group or ring considered there is a constant matrix U such that

$$U A_1(r)\, U^{-1} = A_2(r)\,.$$

A representation is *reducible* if it leaves a subspace of the space R of the representation invariant. If so, there is a transformation U which puts all the matrices into the form

$$\begin{pmatrix} P & 0 \\ R & Q \end{pmatrix}$$

where the submatrix P performs a transformation on the subspace.

If the representation leaves invariant two subspaces which together span the total original space, then the matrices may be written in the form

$$\begin{pmatrix} P & 0 \\ 0 & Q \end{pmatrix}$$

and the representation is called *decomposable* or *completely reducible*. For semisimple Lie-groups PETER and WEYL [8] proved the famous theorem, that *all finite reducible representations are also decomposable*. The proof of these authors contains essentially an integration over the whole (finite) volume of the group ("unitary trick"). As the theorem is a purely algebraic statement concerning Lie-rings alone, a proof was desirable which only uses algebraic properties of the Lie-ring.

For the three dimensional rotation group (for which the quadratic form F, defined by (15) can be simply written $L_x^2 + L_y^2 + L_z^2$), such a proof was first given by CASIMIR and VAN DER WAERDEN* [9].

A very elegant, purely algebraic proof for all semisimple Lie-rings was given by BRAUER [10] later on.

---

* Professor VAN DER WAERDEN kindly informed me that the generalisation of this proof to all semisimple groups in this paper contains an error which was first pointed out by FREUDENTHAL.

Brauer's proof is based on his general theorem, which can be formulated as follows for Lie-rings:

Theorem of Brauer: If all representations of a Lie-ring of the particular form

$$\begin{pmatrix} A & 0 \\ C & 0 \end{pmatrix}$$

where A is an irreducible representation, are completely reducible, then all finite representations of the Lie-ring are completely reducible.

We cannot give here the proof of Brauer's theorem. Once it is given, one can use Casimir's quadratic form F to prove the premises of this theorem for all semisimple Lie-rings with the help of the following:

Auxiliary theorem: The eigenvalue λ of Casimir's quadratic form F defined for any semisimple Lie-ring, is different from zero for every non-trivial finite irreducible representation of the Lie-ring.

By the "trivial representation" of the Lie-ring we understand the one in which 0 corresponds to every element. We do not give here the proof of the auxiliary theorem either, which uses the construction of all irreducible representations for all semisimple Lie-algebras by Cartan and Weyl.

Using both Brauer's general theorem and the auxiliary theorem, one can proceed in the following way. Since the form F of Casimir commutes with every element of the Lie-ring, it is (by Schur's well known Lemma, see Weyl [11], p. 83) a multiple of the unit matrix, $\lambda \cdot I$, where according to the auxiliary theorem $\lambda \neq 0$, except for the trivial representation. Hence F can be written

$$F = \begin{pmatrix} \lambda \cdot I & 0 \\ K & 0 \end{pmatrix} .$$

For $\lambda \neq 0$, by application of the transformation

$$T = \begin{pmatrix} I & 0 \\ \dfrac{K}{\lambda} & 0 \end{pmatrix}$$

one obtains

$$TFT^{-1} = \begin{pmatrix} \lambda \cdot I & 0 \\ 0 & 0 \end{pmatrix} .$$

Since the elements of the Lie-ring commute with F, this transformation decomposes them, i. e. the reducible representation is decomposable.

III. Lie algebra of inhomogeneous rotation groups

Such a group consists of the linear transformations on an n-dimensional space which leave invariant forms of the type

$$\sum_{i=1}^{n} (x_i - y_i)^2$$

where $(x_1, \ldots, x_n)$, $(y_1, \ldots, y_n)$ are the coordinates of points in this space. The typical transformation is thus

$$x_i \to x_i' = \sum_{j=1}^{n} x_j \, a_{ji} + \sum_{i=1}^{n} b_i$$

where the coefficients a_{ji} form an orthogonal matrix; the coefficients a_{ji} and b_i are usually supposed to be real.

Thus this group includes *translations* as well as rotations; it is not semisimple, for the translations form an invariant Abelian subgroup. The infinitesimal operators are obtained by consideration of the appropriate Taylor series associated with functions on the n-dimensional space. We obtain:

$$e_{\lambda\mu} \equiv x_\lambda \frac{\partial}{\partial x_\mu} - x_\mu \frac{\partial}{\partial x_\lambda} \; ; \quad d_\lambda \equiv \frac{\partial}{\partial x_\lambda} \, .$$

It is sufficient to use this special representation of the Lie-algebra in order to derive its commutation relations:

i) $[e_{\lambda\mu}, e_{\varrho\sigma}] = \delta_{\lambda\sigma} e_{\mu\varrho} + \delta_{\mu\varrho} e_{\lambda\sigma} - \delta_{\lambda\varrho} e_{\mu\sigma} - \delta_{\mu\sigma} e_{\lambda\varrho} \, ,$

ii) $[d_\lambda, e_{\mu\nu}] = \delta_{\lambda\mu} d_\nu - \delta_{\lambda\nu} d_\mu \, ,$

iii) $[d_\lambda, d_\mu] = 0 \, .$

The Lie algebra for the *homogeneous* rotation group is given by omitting the d_λ.

We write conventionally $e_{\mu\nu} \equiv -e_{\nu\mu} \equiv iJ_{\mu\nu}$; $d_\mu \equiv ip_\mu$. $J_{\mu\nu}$ and p_μ are then *Hermitian* operators.

Representations of the Lie algebra of the three-dimensional inhomogeneous rotation group

In the case of $n = 3$ we also write $J_{12} = J_3$, $J_{23} = J_1$, $J_{31} = J_2$ and $J_1 \pm iJ_2 = J_\pm$, $p_1 \pm ip_2 = p_\pm$ (J_3 and p_3 remain).

The commutation relations are thus:

$$[J_3, J_+] = J_+; \quad [J_3, J_-] = -J_-; \quad [J_+, J_-] = 2J_3;$$
$$[J_+, p_-] = [J_-, p_+] = 2p_3; \quad [J_3, p_\pm] = [p_3, J_\pm] = \pm p_\pm \, .$$

Other commutator brackets are zero.

There are two *invariants* (i. e. quantities which commute with all 6 operators), which may be constructed from the infinitesimal operators, namely

$$\vec{p}^2 = p_1^2 + p_2^2 + p_3^2 \quad \text{and} \quad \vec{p} \cdot \vec{J} = \vec{J} \cdot \vec{p} = p_1 J_1 + p_2 J_2 + p_3 J_3 \, .$$

$\vec{J}^2 = J_1^2 + J_2^2 + J_3^2$ is not an invariant of the inhomogeneous group. We now consider the value of $[\vec{J}^2, p_j]$. Take for example

$$[\vec{J}^2, p_1] = J_2[J_2, p_1] + [J_2, p_1] J_2 + J_3[J_3, p_1] + [J_3, p_1] J_3$$
$$= +i[-J_2 p_3 - p_3 J_2 + J_3 p_2 + p_2 J_3] = 2i[p_2 J_3 - J_2 p_3] \, .$$

Similar results of course will be obtained for $[\vec{J}^2, p_2]$ and $[\vec{J}^2, p_3]$. We obtain in a similar way $[\vec{J}^2, [\vec{J}^2, p_j]] = -4(\vec{J} \cdot \vec{p}) J_j + 2(\vec{J}^2 p_j + p_j \vec{J}^2)$, a relation which will be employed in the investigation of the representations of the Lie algebra.

We shall assume J_3 is represented by a diagonal matrix:

$$(m |J_3| m') = m \, \delta_{mm'} \, .$$

The commutation rules give

$$(m' |J_\pm| m'') \neq 0 \quad \text{only for} \quad m'' = m \mp 1 \, ,$$

i. e. the m values in a given representation differ by integers. We see also that $J_+ J_-$ is diagonal in m.

Now let us write $\varphi(m)$ as the (m, m) matrix element of $J_+ J_-$

$$(m |J_+| m - 1) (m - 1 |J_-| m) \equiv \varphi(m)$$

and obtain from the commutation relations

$$\varphi(m) - \varphi(m + 1) = 2m \, ; \quad \text{i. e.} \quad \varphi(m) = \text{constant} - m(m - 1) \, .$$

Now

$$\vec{J}^2 = \frac{1}{2} (J_+ J_- + J_- J_+) + J_3^2 \, .$$

The eigenvalue of $\vec{J}^2$ in this representation is therefore

$$\frac{1}{2} (\varphi(m) + \varphi(m + 1)) + m^2$$

since $(m |J_- J_+| m) = (m |J_-| m + 1) (m + 1 |J_+| m) = \varphi(m + 1)$. Now assume the representation is of finite degree; the trace of the commutator $[J_+, J_-]$ is then zero, i. e. $\Sigma m = 0$ which implies the m values are either integer or half-odd-integer. The finiteness of the representation may be proved from the assumption that it is Hermitian (i. e. that $\varphi(m) \geq 0$). We obtain in either case

$$-j \leq m \leq j \, ; \quad \varphi(-j) = \varphi(j + 1) = 0$$

and the above constant $= j(j + 1) = $ eigenvalue of $\vec{J}^2$.

The above discussion allows us to write down the matrices of J_3, J_+ and J_- in the j, m representation. They may be taken with the usual choice of an arbitrary phase, as

$$(j m |J_3| j m) = m \, ,$$
$$(j m |J_+| j m - 1) = \sqrt{(j + m) (j - m + 1)} \, ,$$
$$(j m |J_-| j m + 1) = \sqrt{(j - m) (j + m + 1)} \, .$$

Let us now consider the representation of the p_i; no reference will be made at this stage to the fact that the p_i commute with each other; the theory developed will thus be applicable to the representation of the homogeneous rotation group in four dimensions. The commutation relations with the p_i then show that the non-zero matrix elements of the p_i may be factorized, the left-hand components being independent of m (cf. Weyl [12], p. 200). To determine the dependence on m of the matrix element of p_-, we refer to the relation $[p_-, J_-] = 0$. For example, if

$$j' = j$$

we have

$$\frac{(jm\,|p_-|\,j\,m+1)}{[(j-m)\,(j+m+1)]^{\frac{1}{2}}} = \frac{(j\,m-1\,|p_-|\,jm)}{[(j-m+1)\,(j+m)]^{\frac{1}{2}}}$$

i. e. each ratio is independent of m, and we may write

$$(jm\,|p_-|\,j\,m+1) = (j\,|p_-|\,j)\,\sqrt{(j-m)\,(j+m+1)}\ .$$

The dependence of the matrix of p_3 on m is determined by use of the relation $[J_+, p_-] = 2p_3$, and that of p_+ by reference to the Hermitian nature of the representation.

We obtain thus:

$$
\begin{aligned}
(jm\,|p_3|\,j\,m) &= (j\,|p|\,j)\,m\,,\\
(jm\,|p_+|\,j\,m-1) &= (j\,|p|\,j)\,\sqrt{(j+m)\,(j-m+1)}\,,\\
(jm\,|p_-|\,j\,m+1) &= (j\,|p|\,j)\,\sqrt{(j-m)\,(j+m+1)}\,,\\
(jm\,|p_3|\,j+1\,m) &= (j\,|p|\,j+1)\,\sqrt{(j+m+1)\,(j-m+1)}\,,\\
(jm\,|p_+|\,j+1\,m-1) &= (j\,|p|\,j+1)\,\sqrt{(j-m+2)\,(j-m+1)}\,,\\
(jm\,|p_-|\,j+1\,m+1) &= -(j\,|p|\,j+1)\,\sqrt{(j+m+2)\,(j+m+1)}\,,
\end{aligned}
$$

$$
\left.
\begin{aligned}
(jm\,|p_3|\,j-1\,m) &= (j\,|p|\,j-1)\,\sqrt{(j+m)\,(j-m)}\\
(jm\,|p_+|\,j-1\,m-1) &= -(j\,|p|\,j-1)\,\sqrt{(j+m)\,(j+m-1)}\\
(jm\,|p_-|\,j-1\,m+1) &= (j\,|p|\,j-1)\,\sqrt{(j-m)\,(j-m-1)}
\end{aligned}
\right\}
\begin{aligned}
&\text{N. B.}\\
&(j\,|p|\,j-1)=0\\
&\text{if } j=0, \tfrac{1}{2}\,.
\end{aligned}
$$

In particular the diagonal matrix element of $p_+p_- - p_-p_+$ is:

$$(jm\,|p_+p_- - p_-p_+|\,jm) = 2m\{-|(j\,|p|\,j+1)|^2\,(2j+3) +$$
$$+ |(j\,|p|\,j-1)|^2\,(2j-1) + |(j\,|p|\,j)|^2\}\,.$$

Let us *define*

$$(j\,|p|\,j+1)\,(j+1\,|p|\,j)\,(2j+3)\,(2j+1) \equiv \varphi(j)\,.$$

Then

$$(j\,|p|\,j-1)\,(j-1\,|p|\,j)\,(2j+1)\,(2j-1) = \varphi(j-1)$$
$$\{\varphi(-1) = 0\}$$

and

$$(jm\,|p_+p_- - p_-p_+|\,jm) = 2m\left\{\frac{-\varphi(j) + \varphi(j+1)}{2j+1} + (j\,|p|\,j)^2\right\}.$$

The invariant

$$\vec{p}^2 = p_3^2 + \frac{1}{2}\,(p_+p_- + p_-p_+)$$

has the eigenvalue

$$p^2 = \left\{\varphi(j)\cdot\frac{j+1}{2j+1} + \varphi(j-1)\,\frac{j}{2j+1} + |(j\,|p|\,j)|^2\,j(j+1)\right\}$$

and the invariant

$$\vec{J}\cdot\vec{p} = \vec{p}\cdot\vec{J}$$

the eigenvalue
$$(j \, |p| \, j) \cdot j \, (j + 1) \, .$$
These two quantities are constants for a given representation. Now $4 (\vec{J} \cdot \vec{p}) \vec{J}$ has clearly non-zero matrix elements only when they are diagonal in j. It follows from this and the relation

that
$$[\vec{J}^2, [\vec{J}^2, \vec{p}]] = - 4 (\vec{J} \cdot \vec{p}) \vec{J} + 2 (\vec{J}^2 \vec{p} + \vec{p} \vec{J}^2)$$

$$(j' m' \, |(\vec{J}^2)^2 \vec{p} - 2 \vec{J}^2 \vec{p} \vec{J}^2 + \vec{p} (\vec{J}^2)^2 - 2 (\vec{J}^2 \vec{p} + \vec{p} \vec{J}^2)| \, j'' \, m'')$$
is zero for $j' \neq j''$.

If $(j' \, |p| \, j'') \neq 0$ we have then
$$(j' + j'' + 2) \, (j' + j'') \, (j' - j'' + 1) \, (j' - j'' - 1) = 0 \, ,$$
i. e. if $j' \neq j''$ we have the selection rule $j' - j'' = \pm 1$.

We have not up to now made use of the fact that the p's commute with each other. We now take this into account. We distinguish two cases according to whether the eigenvalue
$$(j \, |p| \, j) \, j \, (j + 1) = C$$
of $\vec{J} \cdot \vec{p}$ is zero or not.

a) $C = 0$. The matrix element of $p_+ p_- - p_- p_+ \, (= 0)$ gives
$$\varphi (j) \doteq \varphi (j - 1) = \text{constant} \, .$$
The constant is clearly equal to the eigenvalue p^2 of $\vec{p}^2$. It follows from the definition of $\varphi (j)$ that
$$|(j \, |p| \, j + 1)|^2 = \frac{p^2}{(2j + 3) \, (2j + 1)} \, .$$
Now if the lowest possible value of j, namely j_0, were non-zero, we would have a contradiction in the relations
$$0 = 2m \left(\frac{- \varphi (j) + \varphi (j - 1)}{2j + 1} \right)$$
if we suppose $p^2 \neq 0$ (i. e. due to the supposed $\varphi (j_0 - 1) = 0$). If we take $j_0 = 0$ we have in this case $m = 0$ also, avoiding the contradiction. Hence we have an infinite representation with j taking all values from zero.

b) $C \neq 0$. The matrix element of $p_+ p_- - p_- p_+$ now gives
$$\frac{- \varphi (j) + \varphi (j - 1)}{2j + 1} + \frac{C^2}{j^2 (j + 1)^2} = 0 \, ;$$
i. e.
$$\varphi (j) - \varphi (j - 1) = C^2 \left(\frac{1}{j^2} - \frac{1}{(j + 1)^2} \right) \, .$$
Hence $\varphi (j) + \dfrac{C^2}{(j + 1)^2} = \text{constant}$.

Substitution into the expression for the eigenvalue p^2 of $\vec{p}^2$ gives
$$\varphi (j) \frac{j + 1}{2j + 1} + \left[\varphi (j) - C^2 \left(\frac{1}{j^2} - \frac{1}{(j + 1)^2} \right) \right] \frac{j}{2j + 1} +$$
$$+ \frac{C^2}{j (j + 1)} = \varphi (j) + \frac{C^2}{(j + 1)^2} \, .$$
i. e. the constant just mentioned equals p^2.

We now determine the minimum value of j, namely j_0, by setting $\varphi(j_0 - 1) = 0$. Then

$$C^2 = j_0^2 p^2 \quad \text{and} \quad C = \pm \sqrt{p^2} \cdot j_0 .$$

We obtain therefore two non-equivalent representations corresponding to the eigenvalues of $\vec{J} \cdot \vec{p}$ having the two values $\pm \sqrt{p^2} \cdot j_0$. Now j_0 is the component of $\vec{J}$ parallel to $\vec{p}$

$$\pm j_0 = \frac{\vec{J} \cdot \vec{p}}{\sqrt{p^2}}$$

and can be integer or half-odd-integer. If we consider the group which includes reflections, the representation includes both signs of C, i. e. the reflection reverses the sign of $\vec{J} \cdot \vec{p}$.

IV. The four-dimensional rotation group

We consider first the Lie algebra for the proper orthogonal group, with the invariant form $x_1^2 + x_2^2 + x_3^2 + x_4^2$. We write down the commutation relations for the 6 infinitesimal operators

$$J_{\lambda\mu} = i \left\{ x_\mu \frac{\partial}{\partial x_\lambda} - x_\lambda \frac{\partial}{\partial x_\mu} \right\}$$

which we rewrite for convenience in the following way:

$$J_{23} \to M_1, \quad J_{31} \to M_2, \quad J_{12} \to M_3,$$
$$J_{41} \to N_1, \quad J_{42} \to N_2, \quad J_{43} \to N_3.$$

Note that the M_i have the same commutation relations as the infinitesimal operators of the homogeneous 3-dimensional orthogonal group.

$$[M_1, M_2] = iM_3, \ [M_2 M_3] = iM_1, \ [M_3, M_1] = iM_2,$$
$$[M_1, N_2] = [N_1, M_2] = iN_3,$$
$$[M_2, N_3] = [N_2, M_3] = iN_1,$$
$$[M_3, N_1] = [N_3, M_1] = iN_2,$$
$$[M_1, N_1] = [M_2, N_2] = [M_3, N_3] = 0,$$
$$[N_1, N_2] = iM_3, \ [N_2, N_3] = iM_1, \ [N_3, N_1] = iM_2.$$

We see further that the relations between the M's and the N's are the same as those between the J's and p's in the case of the inhomogeneous 3-dimensional rotation group, whose representation we have already considered. The only difference is that the N's do not commute among themselves as do the components of $\vec{p}$.

We have again two invariants

$$F = \frac{1}{2} \sum_{\mu < \nu} J_{\mu\nu}^2 = \frac{1}{2} (\vec{M}^2 + \vec{N}^2)$$

$$G = J_{41} J_{23} + J_{42} J_{31} + J_{43} J_{12} = \vec{M} \cdot \vec{N} = \vec{N} \cdot \vec{M}$$

[regarding the triples $(M_1 \, M_2 \, M_3)$ and $(N_1 \, N_2 \, N_3)$ as sets of vector components].

A certain linear combination of the infinitesimal operators is of interest

$$\vec{K} = \tfrac{1}{2}\,(\vec{M} + \vec{N})\,; \quad \vec{L} = \tfrac{1}{2}\,(\vec{M} - \vec{N})\,.$$

The components of $\vec{K}$ and $\vec{L}$ commute with each other; i. e.

$$[K_i, L_j] = 0 \quad \text{for all} \quad i, j\,.$$

In addition

$$[K_1, K_2] = i\,K_3 \text{ etc.}$$
$$[L_1, L_2] = i\,L_3 \text{ etc.}$$

and

$$F = \vec{K}^2 + \vec{L}^2\,, \quad G = \vec{K}^2 - \vec{L}^2\,.$$

This shows that the four-dimensional rotation group may be considered as the direct product of two 3-dimensional rotation groups. This relation is only valid for the proper rotation group and not for the Lorentz group. If we take $\vec{K}$ and $\vec{L}$ as Hermitian operators we get, using the known theory of the homogeneous 3-dimensional group, a set of finite representations of the 4-dimensional group, each representation with

$$\vec{K}^2 = k\,(k+1)\,, \quad \vec{L}^2 = l\,(l+1)$$

and degree $(2k+1)\,(2l+1)$. Each representation may be labelled by the number pair (k, l) and the invariants F and G have the values

$$F = k\,(k+1) + l\,(l+1)\,,$$
$$G = k\,(k+1) - l\,(l+1) = (k-l)\,(k+l+1)\,.$$

The mirror representations (k, l) and (l, k) correspond to a change of sign of $\vec{N}$, i. e. to a reflection of the spatial coordinates.

We shall now adapt the representation theory of the inhomogeneous 3-dimensional rotation group to give that of the homogeneous 4-dimensional group. We start by making $\vec{M}^2$, $\vec{N}^2$ and M_3 diagonal; we know already that the eigenvalues of $\vec{M}^2$ and M_3 are $j\,(j+1)$ and m respectively, where $-j \leqq m \leqq j$. From the previous theory, in analogy with the treatment of $(p_+ p_- - p_- p_+)$,

$$\langle j\,m \,|N_+ N_- - N_- N_+|\, j\,m\rangle = 2m\{-\langle j\,|N|\,j+1\rangle\,\langle j+1\,|N|\,j\rangle\,(2j+3) +$$
$$+\,\langle j\,|N|\,j-1\rangle\,\langle j-1\,|N|\,j\rangle\,(2j-1) + \langle j\,|N|\,j\rangle^2\}\,,$$
$$\langle j\,m\,|N^2|\,j\,m\rangle = \{\langle j\,|N|\,j+1\rangle\,\langle j+1\,|N|\,j\rangle\,(2j+3)\,(j+1) +$$
$$+\,\langle j\,|N|\,j-1\rangle\,\langle j-1\,|N|\,j\rangle\,(2j-1)\,j + \langle j\,|N|\,j\rangle^2\,j\,(j+1)\}\,.$$

We write

$$\langle j\,|N|\,j-1\rangle\,\langle j-1\,|N|\,j\rangle\,(2j+1)\,(2j-1) \equiv \varphi(j)$$

hence

$$\langle j\,|N|\,j+1\rangle\,\langle j+1\,|N|\,j\rangle\,(2j+1)\,(2j+3) = \varphi(j+1)\,.$$

On evaluating $G = \vec{M} \cdot \vec{N}$ we get $G = \langle j \,|N|\, j \rangle j (j + 1)$. Now $N_+ N_- - N_- N_+ = 2 M_3$; hence we may rewrite the above equations

$$\varphi(j + 1) - \varphi(j) = -(2j + 1) + \frac{G^2 (2j + 1)}{j^2 (j + 1)^2} \quad \text{when } j \neq 0,$$

$$2F = \vec{M}^2 + \vec{N}^2 = \varphi(j + 1) \left(\frac{j + 1}{2j + 1} \right) + \varphi(j) \left(\frac{j}{j(j + 1)} \right) + \frac{G^2}{j(j + 1)} + j(j + 1).$$

The first of the new equations gives us

$$\varphi(j) = \text{constant} - \frac{G^2}{j^2} - j^2$$

and putting this result into the second equation gives

$$\text{constant} = 2F + 1.$$

We discuss first the finite representations; for the proper orthogonal group these are also Hermitian.

There are two cases:

a) $G = 0$: a maximum $j = n$ must exist, i. e. $\varphi(n + 1) = 0$. Hence

$$\varphi(j) = 2F + 1 - j^2; \quad 2F + 1 = (n + 1)^2;$$

thus

$$0 \leq j \leq n$$

(we get no boundary condition for $j = 0$).

b) $G \neq 0$: now we get a minimum $j \equiv j_0 \neq 0$; i. e. $j_0 \leq j \leq n$. Then

$$\varphi(j_0) = \varphi(n + 1) = 0,$$

therefore

$$\frac{G^2}{(n + 1)^2} + (n + 1)^2 = \frac{G^2}{j_0^2} + j_0^2 = 2F + 1$$

and

$$G^2 \left(\frac{1}{j_0^2} - \frac{1}{(n + 1)^2} \right) = (n + 1)^2 - j_0^2;$$

i. e.

$$G^2 = j_0^2 (n + 1)^2.$$

Hence

$$2F = (n + 1)^2 + j_0^2 - 1 = n(n + 2) + j_0^2.$$

The degree of the representation is

$$\sum_{j = j_0}^{n} (2j + 1) = (n + 1)^2 - j_0^2 = (n + 1 + j_0)(n + 1 - j_0).$$

The representation just found must be equivalent to one of those already derived; the degrees and the values of the invariants must coincide, hence we get $n = k + l; j_0 = |k - l|$, i. e.

a) if $G > 0$ $\quad k = \frac{1}{2}(n + j_0); \quad l = \frac{1}{2}(n - j_0),$

b) if $G < 0$ $\quad k = \frac{1}{2}(n - j_0); \quad l = \frac{1}{2}(n + j_0),$

c) if $G = 0$ $\quad j_0 = 0$ and $k = l = \frac{1}{2}n.$

Thus the mirror representations have different signs of G.

The representations just derived are also representations of the Lorentz group; they are, however, not in this case Hermitian representations; we shall see that the Hermitian representations are of infinite dimension.

V. The Lorentz group*

The Lorentz transformations differ from the orthogonal transformations by the reality conditions which can be expressed by putting $x_4 = i \cdot x_0$ so that now x_0 is real instead of x_4. The Lorentz transformations therefore leave the indefinite quadratic form $-x_0^2 + x_1^2 + x_2^2 + x_3^2$ of the real variables x_0, x_1, x_2, x_3 invariant. Correspondingly the Hermitian infinitesimal operators of the Lorentz group are given by the unchanged J_{23}, J_{31}, J_{12} and

$$J_{0k} = \frac{1}{i} \left(x_0 \frac{\partial}{\partial x_k} + x_k \frac{\partial}{\partial x_0} \right) = \frac{1}{i} J_{4k} (k = 1, 2, 3) .$$

In the following we substitute therefore always $i \cdot \vec{N}$ for $\vec{N}$ and correspondingly $i \cdot G$ for the earlier invariant G, putting

$$\vec{M} = (J_{23}, J_{31}, J_{12}) ,$$
$$\vec{N} = (J_{01}, J_{02}, J_{03}) .$$

Therefore we now obtain the new commutation rules

$$[N_1, N_2] = -i M_3, \ldots,$$

the other commutation rules remaining unchanged, and the invariants

$$F = \frac{1}{2} (M^2 - N^2) ,$$
$$G = \vec{M} \vec{N} .$$

We now define

$$\psi(j) = (j |N| j - 1) (j - 1 |N| j) (2j - 1) (2j + 1)$$

instead of the earlier $\varphi(j)$. By changing the appropriate signs in the expression obtained before for $\varphi(j)$, we have now

$$\psi(j) = -(2F + 1) - \frac{G^2}{j^2} + j^2 .$$

We now consider the possible Hermitian representations. We shall find that they are of infinite dimension, for there is now no upper limit on the value of j due to the requirement that $\psi(j)$ takes non-negative values.

a) $G \neq 0$. There must clearly be a minimum non-zero value of j since we recall that $G = j (j + 1) \langle j |N| j \rangle$. We call this minimum j-value j_0; and we see from the definition of $\psi(j)$ that

$$\psi(j_0) = 0 .$$

---

* See also [13—17].

Hence

$$2F = j_0^2 - 1 - \frac{G^2}{j_0^2} \, ,$$

and

$$\psi(j) = j^2 - j_0^2 + G^2 \left(\frac{1}{j_0^2} - \frac{1}{j^2} \right)$$

which is non-negative for all $j \geqq j_0$.

b) $G = 0$, $j \geqq j_0 > 0$. We have again $\psi(j_0) = 0$, and hence

$$2F + 1 = j_0^2 \, ; \quad \psi(j) = j^2 - j_0^2 \, .$$

c) $G = 0$, $j \geqq 0$. This case is particularly interesting as here the vanishing of $\psi(0)$ is not required but only $\psi(j) > 0$ for $j = 1, 2, \ldots,$ which means $F < 0$.

A different partition of these cases is obtained by putting

1) Principal series

$$2F = j_0^2 - 1 - \nu^2 \, ; \quad G = j_0 \cdot \nu \, ,$$

$$j_0 = 0, 1, 2, \ldots \quad \text{or} \quad {}^1/_2, {}^3/_2, \ldots ; \quad \nu \text{ real} \, .$$

2) Complementary series or critical strip

$$2F = -1 + \alpha^2 \, ; \quad G = j_0 = 0 \, ,$$

$$0 < \alpha < 1 \, .$$

A part of case c) is contained in 1), another part in 2), while the cases a) and b) are contained in 1). It turns out that due to the infinite volume of the group space of the Lorentz group only the part 1) of the irreducible representations is needed to form a complete orthogonal set. The other part, 2), is not used in physics.

The inhomogeneous Lorentz group

We must add the infinitesimal operators of translations; these involve the commutation relations

$$\left. \begin{array}{l} [p_\lambda, J_{\mu\nu}] = i(\delta_{\lambda\nu} p_\mu - \delta_{\lambda\mu} p_\nu) \\ [p_\mu, p_\nu] = 0 \end{array} \right\} \lambda, \mu, \nu = 1, 2, 3, 4 \, .$$

Since we chose previously to put $i\, x_0 = x_4$ we now have $+i\, p_0 = p_4$. Thus we have the original operators with their original commutation relations from the 3-dimensional inhomogeneous group, namely J_1, J_2, J_3 and p_1, p_2, p_3 plus the operators p_0 and N_1, N_2, and N_3. The new commutation relations involving p_0 and N_1, N_2, N_3 are thus

$$[p_1, N_1] = +i\, p_0 \ldots \ldots \qquad [p_0, N_1] = +i\, p_1 \ldots \ldots$$

$$[p_2, N_1] = [p_1, N_2] = \cdots \cdots = 0 \quad [p_0, M_1] = \cdots = 0 \, .$$

F and G are no longer invariants; we now have as invariants the square of the four-momentum

$$P = -p_\nu p_\nu = p_0^2 - \vec{p}^2 \, ,$$

and another invariant which we construct by first introducing the skew-symmetric tensor

$$v_{k\lambda\mu} = p_k J_{\lambda\mu} + p_\lambda J_{\mu k} + p_\mu J_{k\lambda} = J_{\lambda\mu} p_k + J_{\mu k} p_\lambda + J_{k\lambda} p_\mu .$$

We define the four-vector

$$(w_1, w_2, w_3, w_4) = -i(v_{234}, v_{314}, v_{124}, v_{321}) .$$

We have now the relations

$$[w_\lambda, J_{\mu\nu}] = i(\delta_{\lambda\nu} w_\mu - \delta_{\lambda\mu} w_\nu) ;$$
$$[w_\lambda, p_\mu] = 0 ;$$
$$[w_\mu, w_\nu] \neq 0 \quad \text{for} \quad \mu \neq \nu ;$$
$$w_\nu p_\nu \equiv 0 .$$

The second invariant is thus

$$W = w_\nu w_\nu = -\frac{1}{6} v_{k\lambda\mu} v_{k\lambda\mu} = \frac{1}{2} (p_\lambda p_\lambda)(J_{\mu\nu} J_{\mu\nu}) - J_{k\mu} J_{k\nu} p_\mu p_\nu .$$

We consider the significance of these two invariants in the rest system. For the first we have $\vec{p} = 0$, $p_4 = im$. Hence $p = m^2$.

For the second, we have, since $\vec{p} = 0$,

$$(w_1, w_2, w_3) = im(J_{23}, J_{31}, J_{12}) \quad \text{and} \quad w_4 = 0 .$$

Hence $\vec{w} = im \times$ (angular momentum in rest system)

$$\text{i. e. } -W = m^2 s(s+1) \quad \text{and} \quad -\frac{W}{P} = s(s+1) ,$$

where s is the spin, provided that $m \neq 0$. There are clearly $2s + 1$ independent states for a given momentum vector. The cases $s = 0$, $\frac{1}{2}$ and 1 correspond to the Klein-Gordon, Dirac and Proca equations, respectively.

If $P = 0$, there are two cases:

a) $W = 0$. We have $w_\nu = \lambda p_\nu$ from $w_\nu p_\nu = 0$. Here λ is essentially the spin s. To show this we make use of the expressions in the 3-dimensional vector operators

$$\vec{w} = p_0 \vec{M} - (\vec{p} \times \vec{N}) = \lambda \vec{p} ,$$
$$w_0 = \vec{p} \cdot \vec{M} = \lambda p_0 .$$

Since $p_0 = \pm \sqrt{p_1^2 + p_2^2 + p_3^2} = \pm p$, we have $\lambda = \pm \dfrac{\vec{p} \cdot \vec{M}}{p}$.

The representation theory of the 3-dimensional inhomogeneous group shows us that the magnitude of this quantity $\Big($referred to there as

$j_0 = \pm \dfrac{\vec{J} \cdot \vec{p}}{p}\Big)$ is the minimum j value possible in the given representation; for a given momentum vector there are thus 2 independent states if $\lambda \neq 0$, corresponding to two different states of polarization, and one state when $\lambda = 0$. The cases $s = 0$, $\frac{1}{2}$ and 1 correspond to the scalar wave equation, the neutrino equation, and Maxwell's equations respectively.

In the particular case $W = P = 0$ considered here the representation of the 3-dimensional inhomogeneous rotation group can easily be supplemented to a representation of the inhomogeneous Lorentz group. (Here we are dealing with their infinitesimal transformations only.)

From the commutation rules of the p_i with the N_k follows

$$[\vec{p}, (\vec{p}\,\vec{N})] = i\,p_0\,\vec{p}$$

and with $p_0 = \pm p$,

$$[p, \vec{N}] = \pm i\,\vec{p}.$$

Hence for an arbitrary function $f(p)$ of the absolute value p

$$[f(p), \vec{N}] = \pm i\,f'(p) \cdot \vec{p}$$

and

$$\left[\frac{1}{p}, \vec{N}\right] = \pm i\left(-\frac{1}{p^2}\right) \cdot \vec{p}.$$

The unit vector $\vec{n} = \dfrac{\vec{p}}{p}$ has therefore the commutation relations

$$[n_1, N_1] = p_1\left[\frac{1}{p}, N_1\right] + [p_1, N_1]\frac{1}{p} = \pm i(1 - n_1^2), \ldots,$$

$$[n_1, N_2] = p_1\left[\frac{1}{p}, N_2\right] = \mp i\,n_1\,n_2, \ldots.$$

Therefore, with

$$N_{||} \equiv \frac{1}{2}\,[(\vec{n}\,\vec{N}) + (\vec{N}\,\vec{n})] = (\vec{n}\,\vec{N}) \mp i, \quad [n_i, N_{||}] = 0,$$

but

$$[p, N_{||}] = \pm i\,p.$$

As

$$0 = (\vec{p} \times \vec{w}) = p^2\{\pm(\vec{n} \times \vec{M}) + \vec{N} - \vec{n}(N_{||} \pm i)\}$$

the quantities $N_{||}$ and p are the only ones with which the quantities $\vec{M}$ and $\vec{n}$ of the 3-dimensional inhomogeneous rotation group have to be supplemented in order to obtain the case $W = P = 0$ of the inhomogeneous Lorentz group. As $\vec{M}$ and $\vec{n}$ commute with p and $N_{||}$ they retain their matrix representation of the 3-dimensional group.

The continuous variable p with the volume element $p \cdot dp$ (obtained from the invariant volume element $\dfrac{d^3 p}{p_0} = p \cdot dp \cdot dM$ by integration over the directions, $d\Omega$) has then to be incorporated (besides the integers, j, m) in the space of the representation of the inhomogeneous Lorentz group.

In agreement with the commutation rule already mentioned,

$$[p, N_{||}] = \pm i\,p$$

the quantity $N_{||}$ has then to be represented by the operator

$$(N_{||})_{\text{op}}\, F(p) = \mp i\,\frac{\partial}{\partial p}\,(pF)$$

which is Hermitian with respect to the volume element $p \cdot dp$.

b) $W \neq 0$. There is a continuous spectrum of W instead of the spin; these representations do not seem to have any physical significance.

For application of this representation theory to physical problems see refs. [18—21].

References

[1] Lie, S., and G. Scheffers: Vorlesungen über kontinuierliche Gruppen. Leipzig 1893

[2] Racah, G.: Group theory and spectroscopy. Princeton 1951. Page 28 in this volume

[3] Pontrjagin, L.: Topological groups. Princeton, University press, 1939 (see also pp. 181 and 334)

[4] Weyl, H.: The structure and representation of continuous groups. Princeton, 1934—1935 (mimeographed notes; see Chapter 2, Pt. 1)

[5, 6] — Theorie der Darstellung kontinuierlicher halb-einfacher Gruppen durch lineare Transformationen. Math. Z. 23, 271—309 (1925); 24, 328—395 (1926). (Reprinted in: Selecta Hermann Weyl, p. 262. Basel: Birkhauser 1956)

[7] Boerner, H.: Darstellung von Gruppen. Berlin-Göttingen-Heidelberg: Springer 1955

[8] Peter, F., and H. Weyl: Die Vollständigkeit der primitiven Darstellungen einer geschlossenen kontinuierlichen Gruppe. Math. Ann. 97, 737—755 (1927)

[9] Casimir, H., and B. L. van der Waerden: Algebraischer Beweis der vollständigen Reduzibilität der Darstellungen halbeinfacher Liescher Gruppen. Math. Ann. 111, 1—12 (1935)

[10] Brauer, R.: Eine Bedingung für vollständige Reduzibilität von Darstellungen gewöhnlicher und infinitesimaler Gruppen. Math. Z. 41, 330—339 (1936)

[11] Weyl, H.: The classical groups, their invariants and representations. Princeton: University press 1939

[12] — Gruppentheorie und Quantenmechanik. 2. Aufl. Leipzig: Hirzel 1931

[13] Wigner, E.: On unitary representations of the inhomogeneous Lorentz group. Ann. of math. 40, 149—204 (1939)

[14] Harish-Chandra: Infinite irreducible representations of the Lorentz group. Roy. Soc. Proc. A 189, 272—401 (1947)

[15] Bargmann, V., and E. P. Wigner: Group theoretical discussion of relativistic wave equations. Nat. Acad. Sci. Washington, Proc. 34, 211—223 (1948)

[16] — Irreducible unitary representations of the Lorentz group. Ann. Math. 48, 568—640 (1947)

[17] Gelfand, I., and M. Neumark: Unitary representation of the Lorentz group. J. phys. USSR 10, 93—94 (1946)

[18] Pauli, W.: Über das Wasserstoffspektrum vom Standpunkt der neuen Quantenmechanik. Z. Physik 36, 336—363 (1926)

[19] Hulthén, L.: Über die quantenmechanische Herleitung der Balmerterme. Z. Physik 86, 21—23 (1933)

[20] Fock, V.: Zur Theorie des Wasserstoffatoms. Z. Physik 98, 145—154 (1936)

[21] Bargmann, V.: Zur Theorie des Wasserstoffatoms. Z. Physik 99, 576—582 (1936)

Determination of Nuclear Moments with Optical Double Resonance

G. zu Putlitz

I. Physikalisches Institut der
Universität Heidelberg

Received December 1964

Contents

I. Introduction

The knowledge of the static moments and the spins of nuclei is of special interest because their interpretation and calculation has become possible as a result of the development of satisfactory nuclear models [1—4]. A great number of the presently known nuclear magnetic moments* and most of the nuclear quadrupole moments were calculated from the hyperfine structure** of the atomic states [5]. The first experimental determination of hfs splittings resulted with optical interferometric methods which were only then successful when the hfs was larger than the width of the Doppler-broadened spectral line [6]. The first radiofrequency (rf) method was the atomic beam resonance method [7, 8]. It attained a considerably greater precision and resolution and thus enlarged considerably the circle of measurable nuclear moments. The atomic beam method served chiefly for the investigation of atomic ground states but was later also applied to metastable [9, 10] and, in some special cases, to short-lived excited atomic states [11].

In those cases where the atomic ground state exhibits a spherical charge distribution (i. e., an S-state or a state with $J \leq 1/2$) one obtains no information about a nuclear quadrupole moment from the hfs splitting of the state. As a consequence of this the investigation of short-lived excited atomic states with the accuracy of the rf method was desirable. This became possible with introduction of the optical double resonance method by Kastler, Brossel and Bitter [12, 13]. The results of nuclear moment determinations with this double resonance method shall be the topic of this article. First the fundamental features of the method will be described. Then its application in various groups of the periodic table will be discussed in more detail.

II. The optical double resonance method [12, 13]

In the double resonance method the population of the state to be investigated takes place through the absorption of light of a spectral line which was emitted in a transition to the ground state. This light is absorbed by an ensemble of atoms under conditions such that the excited state is not statistically occupied. For clarification consider, for example, the I-spectrum of the even Hg isotopes and a schematic of a double resonance setup (Fig. 1 and 2). The ground state of the HgI spectrum is $^1S_0$; the excited $^3P_1$-state splits up into 3 Zeeman levels in an external magnetic field $H_0 \neq 0$. The light from the light source LS (Fig. 2) — with wavelength $\lambda = 2537 \text{ Å}$ $(6s^2\,^1S_0 - 6s\,6p\,^3P_1)$ — is directed into the resonance chamber R (containing Hg-vapor) by means of a lens and a polarizer P_1 in such a way that the E-vector of the incoming light is parallel to a static magnetic field H_0 which exists at the

---

* Most of the magnetic moment determinations have been made with nuclear resonance methods [5].

** "Hyperfine structure" will be abbreviated with hfs in the following.

chamber. This means that only electric dipole transitions with $\Delta m_J = 0$ can be induced through the resonance absorption of this light. Thus chiefly the $m_J = 0$ Zeeman level of the $^3P_1$-state is populated and the resonance light of the Hg vapor which is observed with the photomultiplier PM is likewise linearly polarized. (The relaxation time of the orientation is essentially larger than the lifetime of the excited Hg state for sufficiently small vapor pressures in the resonance chamber.) An rf-magnetic field H_1 whose frequency equals the Zeeman frequency $\left(\nu_L = g_J \dfrac{\mu_0}{h} \cdot H_0\right.$ where μ_0 = Bohr magneton, h = Planck's constant, g_J = atomic g-factor$\Big)$ is applied to the Hg vapor in the direction perpendicular to H_0. H_1 tends to equalize the occupation of the Zeeman levels of the $^3P_1$-state through the stimulation of magnetic dipole transitions during the lifetime of the excited state. Consequently the resonance light shows, beside the π-component, also the two σ-components.

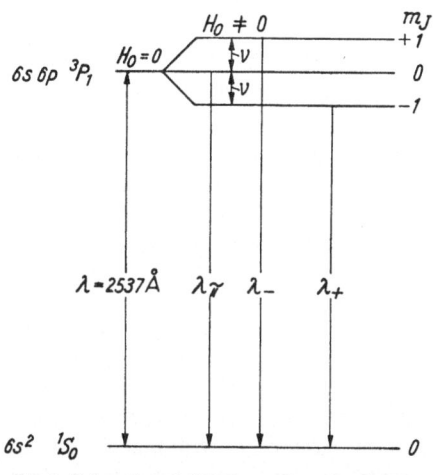

Fig. 1. Relevant part of the term scheme for the even-even Hg isotopes with and without the Zeeman splitting of the excited $^3P_1$-state

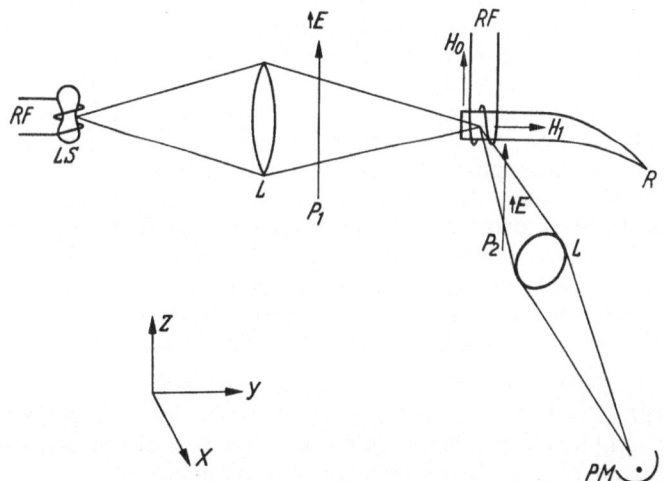

Fig. 2. Schematic diagram of a double resonance apparatus. x-direction: direction of observation of resonance light; y-direction: direction of incoming light; z-direction: direction of static magnetic field H_0; LS: light source; RF: coil of a radio frequency signal generator; L: lens; P: Polarizer with polarization vector $\uparrow E$; R: resonance absorption chamber; H_1: direction of the rf-magnetic field; PM: photomultiplier

This change in the degree of polarization of the resonance light serves for the detection of the rf-transitions in the excited state in that one — by use of a second polarization filter P_2 — allows only π-light to reach the photomultiplier and therefore observes a decrease of the resonance light when H_1 has the correct frequency. One detects, then, transitions in the rf (or microwave) region with quanta in the optical region. Thereby is explained the extreme sensitivity of the method.

Besides this detection method, that of transforming the change in the degree of polarization of the resonance light into an intensity change by means of a polarizer, the angular distribution of the light emitted from the Hg atoms is also usable for the detection. The π-light has an angular distribution of the form $I_\pi = 3 \sin^2\theta$ (θ = angle between H_0 and direction of emission), the σ-light has an angular distribution $I_\sigma = (3/2)(1 + \cos^2\theta)$. The decrease, due to rf transitions, of the re-emitted π-light in the direction perpendicular to H_0 means, then, an increase of the σ-light in the direction of H_0. The detection of this change in the anisotropy of the angular distribution of the resonance light can be carried out with one photomultiplier — which is customarily perpendicular to both H_0 and the direction of the incoming light — as well as with two multipliers, in which case the second registers the light emitted in the direction of H_0. The two signals are then compared by means of a bridge circuit [13]. In both cases it is not necessary to use a polarizer. This is often a great advantage because of the lack of suitable polarizers in the region of relevant wavelengths.

A third possibility for detecting rf transitions in the excited state is based on the circumstance that the reemitted π-light has a different wavelength than the two σ-components of the spectralline. If one makes, in the case of $6p\,^3P_1$ of the even Hg isotopes, the Zeeman splitting of the excited state large in comparison with the Doppler width of the re-emitted spectral line, the intensity distribution of this line shows three completely resolved components λ_-, λ_π, λ_+ (see Fig. 1). In place of the polarizer P_2 a filter which is opaque for the π-component is used. That is in practice an absorption chamber in magnetic field $H = 0$ holding the same kind of atoms as the resonance chamber under conditions which assure the total absorption of this component λ_π. With the stimulation of rf transitions the σ-light with the wave lengths λ_- and λ_+ appears in the resonance radiation and reaches the multiplier with virtually un-weakened intensity, allowing in this way the extremely sensitive detection of changes in occupation within the excited state [14, 15].

The lineshapes of the rf signals were first studied exactly by Brossel and Bitter [13]. In the presently considered case of the even Hg isotopes the signal amplitude S is found to be, as a function of the frequency (ω) and amplitude (H_1) of the applied rf-magnetic field,

$$S(\omega) = k \frac{(\gamma H_1)^2}{(\gamma H_1)^2 + (\omega - \omega_0)^2} \times$$

$$\times \left[\frac{(\gamma H_1)^2}{(1/\tau)^2 + 4(\gamma H_1)^2 + (\omega - \omega_0)^2} + \frac{(\omega - \omega_0)^2}{(1/\tau)^2 + (\gamma H_1)^2 + (\omega - \omega_0)^2} \right] . \tag{1}$$

In this k = constant which depends upon the intensity of the incoming light,

γ = gyromagnetic ratio = $(\mu_0/\hbar) \cdot g_J$,

$\omega = \gamma \cdot H_0$ = resonance frequency,

H_0 = strength of the static magnetic field,

τ = mean life time of the excited state.

The resonance signal is bell-shaped for $\gamma H_1 \ll 1/\tau$ and double-peaked for $\gamma H_1 \gg 1/\tau$. In the latter case the separation of the maxima is $\sqrt{2} \cdot H_1$. Fig. 3 shows the form of the resonance signal for various rf-field strengths.

For small rf-fields it is possible to calculate to good approximation the full width at half maximum of the resonance curve. The result is, under consideration of the first two terms of the expansion,

$$\Delta \omega^2 = (2/\tau)^2 [1 + 5.8 (\gamma H_1 \tau)^2] . \qquad (2)$$

By plotting the square of the line width against the square of the rf-field strength it is thus possible to determine the lifetime of the excited state. This process is often used for the determination of oscillator strengths.

The optical double resonance method (resonance absorption and rf-resonance in the excited atomic state), which was first tried for the $^3P_1$-states of the even Hg isotopes that formed the example above, can likewise be used for the measurement of the hfs of excited atomic states. The relevant hfs levels must simply be populated in a nonstatistical way. This is achieved as in the case of no hfs through a suitable choice of the exciting radiation. With this nonstatistical population of the hfs

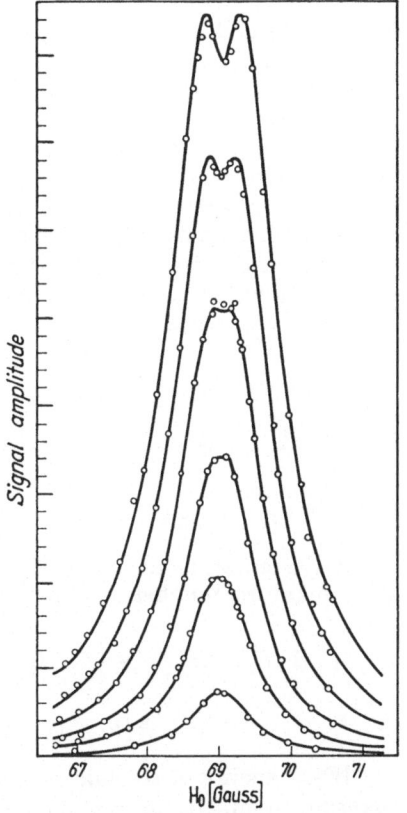

Fig. 3. Line shape of the double resonance signal for various values of the rf-field [13]

levels it is then possible to detect rf transitions between them. Because the transition frequencies are very small compared with the optical frequencies the Doppler broadening of the line plays practically no roll. The sharpness of the signal is determined chiefly by the lifetime of the excited state and amounts to, for example, $\Delta \nu = 3.2$ Mc/s for $\tau = 10^{-7}$ sec and $H_1 \to 0$. Thus this method makes possible a very exact measurement of also such hfs splittings which are completely or almost inaccessible to optical investigation. It offers, furthermore, the advantage that several terms of a spectrum can be subjected to investigation — a possibility which was again considerably extended through the polarized excitation with electron bombardment [16—18].

Quite naturally the first experiments had to do with those atoms whose ground states offered no information about the nuclear quadrupole moment. These are especially the elements which belong to the first and second groups of the periodic table. In fact all measurements of nuclear moments with optical double resonance up to now have occurred in these groups. These atoms fulfil very well the experimental criteria for double resonance investigations: being easily evaporable for resonance absorption and having suitable spectral lines for the population of excited states.

Finally it should be remarked that the double resonance method described above and the related technique of optical pumping are used today for the study of a manifold of physical processes such as the absolute determination of the magnetic moment of the electron [19], relaxation processes and reaction cross sections, quantum electrodynamic corrections in coherence phenomena, and frequency standards. Reference is made to the review papers in Refs. [20—26] since the short description of the method of optical double resonance in this article is by nature only of a qualitative character.

III. Double resonance of excited alkali atoms

1. Hfs of the ^2P-states in external magnetic field $H_0 = 0$

In the external magnetic field $H_0 = 0$ the energy of a hfs state with quantum number F (total angular momentum) is given by [6]

$$W(F) = W_J + A\,\frac{C}{2} + B\,\frac{3/4\,C(C+1) - I(I+1)\,J(J+1)}{2I(2I-1)\,J(2J-1)} \qquad (3)$$

with

$$C = F(F+1) - I(I+1) - J(J+1) \qquad (4)$$

and W_J = energy of the fine structure state, I, J = angular momentum quantum numbers, A = magnetic dipole hfs constant, B = electric quadrupole hfs constant.

For the case of a nuclear spin $I = 3/2$ (Na^{23}, K^{39}, Rb^{87}) the hfs splittings $\nu_{F-F'}$, for a ^2P-state are

$$J = 3/2: \quad \nu_{3-2} = 3A_{3/2} + B$$

$$\nu_{2-1} = 2A_{3/2} - B$$

$$\nu_{1-0} = \quad A_{3/2} - B$$

$$J = 1/2: \quad \nu_{2-1} = 2A_{1/2}.$$

The A-factor is related to the magnetic moment of the nucleus

$$A = \frac{\mu_I \cdot \langle H(0) \rangle}{I \cdot J} \qquad (5a)$$

while the corresponding equation for B is

$$B = e \cdot Q \cdot \langle \varphi_{zz}(0) \rangle \qquad (6a)$$

where μ_I = nuclear magnetic moment, e = electron charge, Q = nuclear electric quadrupole moment, $\langle H(0) \rangle$ = expectation value of the magnetic field at the nucleus produced by the electrons in the state with $m_J = J$, $\langle \varphi_{zz}(0) \rangle$ = expectation value for the state $m_J = J$ of the z-derivative of the z-component of the electric field produced at the nucleus by the electrons.

In the case of the alkali spectra with one valence electron $\langle H(0) \rangle$ and $\langle \varphi_{zz}(0) \rangle$ are simple expressions so that the (5a) and (6a) become (a, b in frequency units)

$$a = \frac{2 \mu_0^2}{h} g_I' \frac{l(l+1)}{j(j+1)} F_{r,j}(1 - \delta)(1 - \varepsilon) \langle r^{-3} \rangle_j \qquad (5b)$$

and

$$b = \frac{e^2 Q}{h} \frac{2j-1}{2j+2} R_{r,j} \langle r^{-3} \rangle_j \qquad (6b)$$

with $g_I' = \frac{\mu_I}{\mu_0 I}$ (nuclear g-factor), l = orbital angular momentum quantum number, δ and ε = corrections for the spatial extent of the charge and the magnetic dipole, respectively, of the nucleus (for the p-electrons of the alkalis $\delta, \varepsilon \approx 0$), $F_{r,j}, R_{r,j}$ = relativistic corrections which are tabulated in Ref. [6], $\langle r^{-3} \rangle_j$ = expectation value of r^{-3} where r is the distance between the nucleus and the valence electron.

In order to determine μ_I and Q from the measured a- and b-factors it is necessary to know $\langle r^{-3} \rangle_j$ very precisely, i. e., to have an exact description of the radial wave function of the valence electron. The calculation of these wave functions is, however, difficult and does not yield very reliable values of $\langle r^{-3} \rangle_j$. Therefore for the case of one electron spectra $\langle r^{-3} \rangle_j$ is found either from the magnetic hfs splitting (if μ_I is known) or from the fine structure splitting which can be expressed in terms of $\langle r^{-3} \rangle_j$ as follows

$$\langle r^{-3} \rangle_j = \frac{\delta \widetilde{W}}{2 \mu_0^2 (l + 1/2) Z_i H_{r,l}} \qquad (7)$$

with $\delta \widetilde{W}$ = fine structure splitting $({}^2P_{1/2} - {}^2P_{3/2})$, $H_{r,l}$ = relativistic correction, Z_i = effective nuclear charge (approximately $Z - 4$ for p-electrons, see Ref. [27] for exacter values).

The values of $\langle r^{-3} \rangle_j$ given by these two methods, by the fine structure splitting and by the a-factor, disagree by up to 10% in the case of the alkalis. For the explanation of this discrepancy one turns today chiefly to the phenomenon of "core polarization", i. e., to the partial excitation of an s-electron of the core into unfilled higher s-shells of the atom. Thus the a-factor is no longer ascribed to the p-electron alone but also contains a contribution which corresponds to that of an s-electron. The size of this contribution can be estimated in the following way. According to Eq. (5b) with $(1 - \delta) \cdot (1 - \varepsilon) \approx 1$ it is possible to write for the a-factors of p-states with the same principal quantum number n

$$\frac{a_{1/2}}{a_{3/2}} = \frac{2l+3}{2l-1} \frac{F_{r,1/2}}{F_{r,3/2}} = 5 \frac{F_{r,1/2}}{F_{r,3/2}} . \tag{8}$$

Rabi and Senitzky [28] found the value $a^{85}_{1/2}/a^{85}_{3/2} = 4.82$ as a result of a study of the $5\,{}^2P_{1/2}$ and $5\,{}^2P_{3/2}$ states of Rb^{85}. The theoretical value with $Z_i = 32$ is 5.45. To take the core polarization into consideration one replaces the measured a-factor by [30]

$$a_j \rightarrow a_j + (g_j - 1)\, \varkappa_s \tag{9}$$

where g_j is the Landé g-factor and where $\varkappa_s$, the contribution due to core polarization, does not depend on j. From the measured values of the a-factors of the two ^2P-states it is then possible to eliminate $\varkappa_s$. In the case of Rb^{85} one obtains $\varkappa_s = 7.2$ Mc/s from the measured values $a^{85}_{1/2} = 120.7(1)$ Mc/s [28] and $a^{85}_{3/2} = 25.029(16)$ Mc/s [31]. One a-factor for the p-electron alone is then $a^{85}_{3/2} = 22.6$ Mc/s, a value which lies closer to the value 23.27 Mc/s calculated from the fine structure splitting than the above measured value does. It is then very desirable to know the a-factor of the $^2P_{1/2}$-state for a calculation of $\langle r^{-3} \rangle_j$ from the magnetic hfs splitting of the $^2P_{3/2}$-state.

When μ_I is known the electric quadrupole moment can be written using Eqs. (5b) and (6b) as

$$Q_{\text{hfs}} = \frac{4}{e^2} \frac{l(l+1)}{j(2j-1)} \mu_0^2 g'_I \frac{F_{r,j}}{R_{r,j}} (1 - \delta)(1 - \varepsilon) \frac{b}{a} . \tag{10}$$

In this formula the corrected (in the manner described above) a-factor is to be used. The Q calculated in this manner is still susceptible to correction via the Sternheimer effect [32–39]. This effect considers the influence of the nuclear quadrupole potential on the electron core and its interaction with the valence electron. The corrected nuclear quadrupole moment is then

$$Q = Q_{\text{hfs}} \left(\frac{1}{1-R} \right) \tag{11}$$

where $(1 - R)^{-1}$ is the Sternheimer correction factor (see part 4 of this section).

2. Double resonance experiments in external magnetic field $H_0 = 0$

The double resonance method allows the direct measurement of the hfs splitting of excited states, i. e., measurement without additional splitting within the hfs F-levels. For this reason most of the experiments on excited states of the alkalis were carried out in external magnetic field $H_0 = 0$. The greater part of these investigations was made by Kopfermann and co-workers. In the following the essential features of these experiments are discussed.

a) Excitation relations and relative intensity of the rf signals

The excitation of the $^2P_{3/2}$-state to be investigated takes place by means of absorption of light of the ground state spectral line $(^2S_{1/2} - {}^2P_{3/2})$.

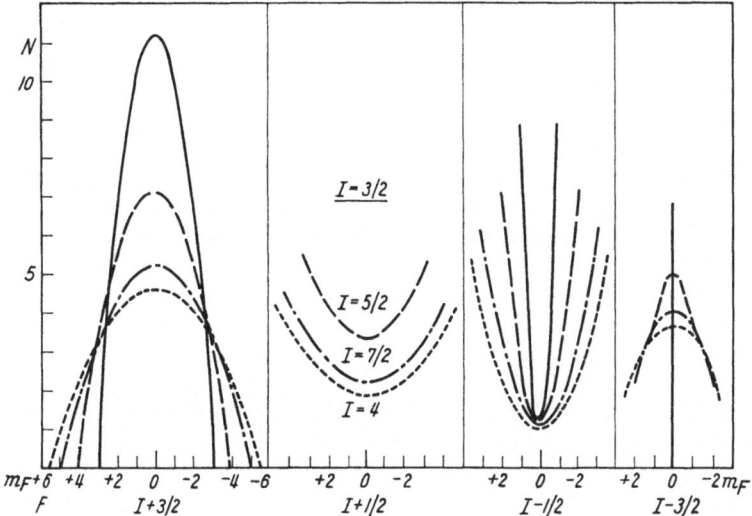

Fig. 4. Graph depicting, for π-excitation, the relative occupation of the individual Zeeman levels of the hyperfine structure of a $^2P_{3/2}$-state for various nuclear spins I

Because the hfs splitting caused by the single p-electron is less than or comparable to the Doppler width of the exciting line, one can assume to good approximation uniform excitation of all hfs levels, i. e., the same energy density for the exciting radiation in the whole absorption region★. The spectral intensity distribution can, if necessary, be influenced by means of Zeeman splitting in a magnetic field. Moreover one must take care so that self reversal of the spectral line does not occur. For the most frequently used type of excitation, linearly polarized π-light, one obtains the occupation numbers depicted in Fig. 4 for the population of the sublevels of the hfs states [40]. These numbers were calculated under the assumption that the axis of quantization coincides with the direction of the

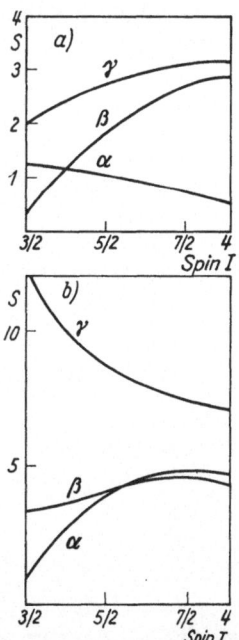

Fig. 5a and b. Relative intensity S of the rf-transitions between the hfs terms of a $^2P_{3/2}$-state for weak (a) and strong (b) rf-field as a function of the nuclear spin. The two cases a, b are not normalized with respect to each other. Population of the $^2P_{3/2}$-state by means of π-excitation. γ: transition from $F = J + I$ to $F = J + I - 1$. β: transition from $F = J + I - 1$ to $F = J + I - 2$; α: transition from $F = J + I - 2$ to $F = J + I - 3$

★ The actual state of affairs is more complicated than this assumption because of the hyperfine structure of the ground state (see Ref. [46]).

incoming light (for a completer discussion of the excitation relations see Refs. [23] and [41]).

With the application of the rf-magnetic field the differences in occupation in the excited levels are partially removed. By observing resonance light emitted perpendicular to the z-axis without the use of a second polarizer in the path of the resonance light one obtains the form [40] depicted in Fig. 5 for the relative strength of the rf transitions $\Delta F = \pm 1$. This is shown for two cases: (a) a weak rf-field in the z-direction and (b) saturation of the rf-field. It is seen that the relative intensities of the various rf transitions between hfs levels of the $^2P_{3/2}$ state are strongly dependent on the nuclear spin I, i. e., on the total angular momentum quantum number F of the hfs levels. A deviation of the experimentally measured intensities from those calculated results from the fact that in the calculation it is assumed that a weak magnetic field in the z-direction is present (i. e., no degeneracy of the m_F-levels). In addition the experimental determination of the relative intensities demands the constancy of the rf-field strength over a large frequency range; this is difficult to attain. Similar considerations for excitation with σ-light yield smaller population differences in the excited state and thus less intense rf signals.

b) Experimental setup

The experimental arrangement for the investigation of excited states of the alkalis has the same general features in all experiments. Thus the description of one apparatus suffices to make these features clear. Fig. 6 shows

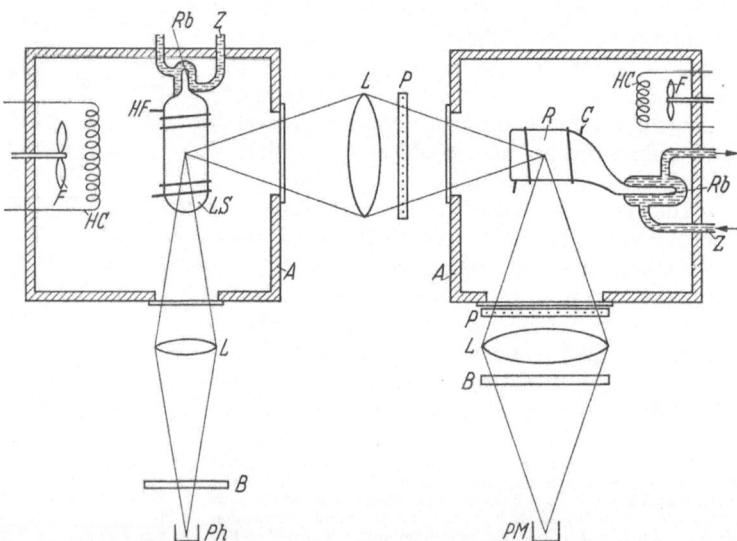

Fig. 6. Schematic of a double resonance apparatus for the alkalis. LS: intense rubidium light source, HF: high frequency coil for electrodeless discharge, L: focussing lenses, P: polarization filter, R: resonance chamber, C: coil for rf-magnetic field, B: blue filter, PM: photomultiplier, Ph: photocell, A: asbestos wall, HC: heating coil, F: fan, Z: oil circulation, Rb: rubidium

such a setup. An electrodeless high frequency discharge in the relevant alkali vapor — with or without some noble gas — serves as the light source. Circulating oil whose stabilized temperature is somewhat lower than that of the lamp container is used to keep the vapor pressure in the lamp constant. Intensity fluctuations of the light source are measured by the use of a suitable filter in combination with a photocell. These measurements are then used to electronically stabilize the intensity of the light source. Via a lens and a polarisation filter the exciting radiation reaches the resonance volume. The solid angle is chosen as large as possible in order to attain intense excitation. However because of the rise in the background of scattered light as a consequence of unfavorable optical image-forming and account of the decrease in the collimation of the exciting radiation (which produces a corresponding decrease in the population differences in the excited states), the size of the solid angle is limited. The resonance light

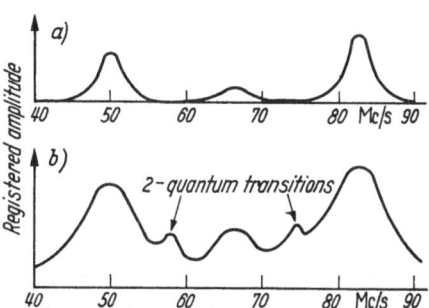

Fig. 7 a and b. Shape of rf-resonance between the hfs levels of the excited $7 \, ^2P_{3/2}$-state of Cs^{133}. a = rf-field about 1 Gauss; b = rf-field about 5 Gauss. The two 2-quantum transitions are visible between the 3 hfs transitions

is observed, with or without the insertion of a polarizer, perpendicular to the incoming light and is measured with a photomultiplier. The resonance cell is found in the tank coil of a generator which produces the rf-magnetic field for the induction of transitions. In order to abstract the signal from the fluctuations of the background, the signal is modulated at a low frequency, amplified with a narrow band amplifier, rectified in a phase-sensitive lock-in circuit, and registered on a graphic recorder.

In carrying out such experiments the following details usually have to be considered. The resonance line is broadened by the hfs-Zeeman effect in the magnetic field of the earth. For this reason the magnetic field of the earth must be compensated out in the region of resonance absorption. It is in general difficult to hold the strength of the rf-field constant over the frequency range covered during the registration of the rf signal. In addition to the measurement of the rf-power with a bolometer the measurement of the oscillator voltage of the generator is often used as a means of determining the relative rf-field strengths. If, because of large rf-field strength, the width of the resonance line is comparable with the hfs splitting, one observes 2-quantum transitions (see Fig. 7) between the resonance signals. The frequency of such a 2-quantum transition is the arithmetic mean of the two resonance signals and the linewidth corresponds to that of the resonance signals at a weaker rf-field [42—44]. For resonances whose width is comparable with the resonance frequency, a frequency shift of the Bloch-Siegert sort [47] can occur. This must, of course, be taken into consideration.

c) Investigation of radioactive alkali isotopes

In a double resonance experiment in which a resonance chamber is used the number of atoms in the vapor phase for a chamber of normal dimensions is about 10^{12}. Consequently it is possible to carry out an experiment with this small number of atoms if reactions with the walls of the chamber do not decrease the number of usable atoms. In the investigation of radioactive Cs isotopes, for example, a workable resonance cell was achieved with 10^{-6} g Cs [45]*. To reduce the loss of Cs

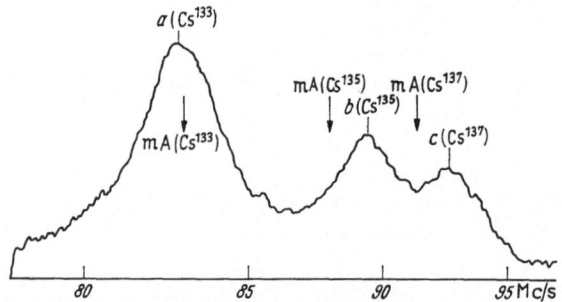

Fig. 8. Rf-resonances in the $7\,^2P_{3/2}$-state of Cs. The three resonances a, b, c correspond to the transition from $F = 5$ to $F = 4$ for the three isotopes Cs¹³³, Cs¹³⁵ and Cs¹³⁷. The arrows with m. A. show the calculated positions of the resonances for $B = 0$

atoms due to wall reactions a 100-fold amount of Rb was used as a buffer. In this way a satisfactory lifetime of the resonance cell was achieved. Fig. 8 shows an experimental curve for the transition $F = 5$ to $F = 4$ of the Cs isotopes 133, 135 and 137 found in the resonance cell. The resonances are not completely resolved because the magnetic moments of the three isotopes differ by only small amounts. An analysis of the resonance structure was nevertheless possible since the magnetic moments of the three isotopes as well as their relative abundances in the cell were precisely known and since the rf-field amplitude was kept constant over the frequency range.

d) Double resonance by means of excitation with hfs components and detection via self-absorption [46]

Through the excitation of hfs levels of the ^2P-state with the hfs components of a spectral line it is possible to achieve a deviation from the statistical occupation in the excited state without using a polarizer. The basis of this technique is that the exciting light is not a continuum but consists rather of two components of different intensity because the hfs splitting of the ground state is large compared to the Doppler width. The exciting component with the larger total intensity is preferentially

---

* In the case of the radioactive Cd isotopes considerably smaller quantities were employed. In contrast to the alkalis, for which a sizable depletion of the atoms on the wall is observed in spite of the use of alkali-resistant glass, Cd does not attack the walls of a quartz cell.

absorbed. Thus the transition probability of this component must be given the greater weight in considering the excitation. Proceeding in this manner BUCKA calculated the occupation ratio $(F = 4):(F = 3)$ = 8.68:7.32, for the excitation of a $^2P_{1/2}$-state of Cs^{133} $(I = 7/2)$ with light of the pertinent ground state transition. For uniform excitation one obtains, on the other hand, the ratio 9 : 7 which is just that of the statistical weights. Rf transitions between the two hfs levels cause an approximation to statistical equilibrium. In the resonance radiation one of the reemitted components appears weaker, the other stronger. This is detected by means of the difference in the self-absorption of the two components in an absorption cell of Cs^{133} located between the resonance cell and the photomultiplier used to measure the resonance light. This method is suited for the study of the $^2P_{3/2}$ – as well as the $^2P_{1/2}$ – state of the alkalis. The latter state can otherwise only be investigated through excitation with circularly polarized radiation.

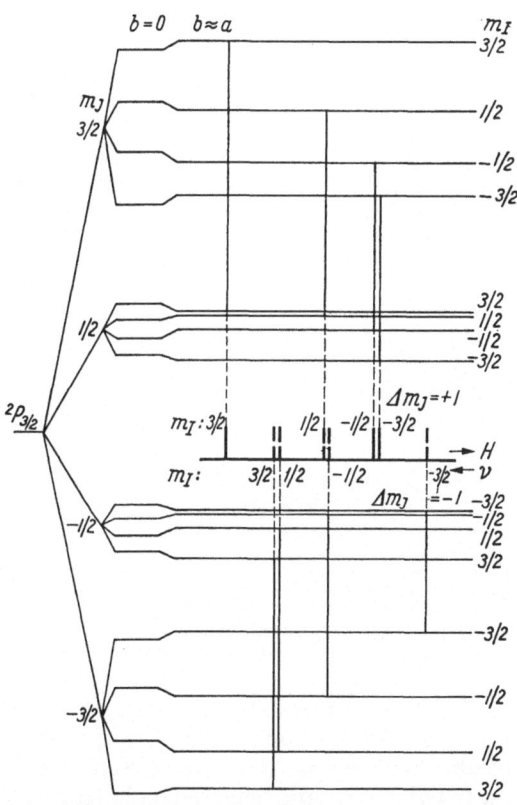

3. Double resonance experiments on excited states of the alkalis in a strong external magnetic field

The magnetic field which the $p_{3/2}$-electrons of the alkalis produce at the nucleus (see Ref. [6]) causes in the case of the sodium and potassium isotopes a hfs splitting which is so small that the double resonance signals between the various hfs levels cannot be completely resolved. In this case the possibility exists of obtaining further information about the A- and B-factors from the hfs of the levels in the

Fig. 9. Energy levels and rf-transitions in a "strong" magnetic field for the 5 $^2P_{3/2}$-state of K^{39} $(I = 3/2)$. The transitions from $m_J = 1/2$ to $m_J = -1/2$ do not contribute to the signal and are therefore not drawn in. In the middle the positions of the 8 observable transitions are depicted on a frequency scale

Paschen-Back region, where I and J are decoupled.

The energy of a Zeeman level of the hfs in a strong magnetic field, where I and J are decoupled, is, neglecting pertubations between hfs levels, given by

$$W_{m_I, m_J} = W_J + \mu_0 \cdot H_0 \cdot m_J \cdot g_J - \mu_0 H_0 m_I \cdot g_I' + A m_I m_J +$$
$$+ \frac{B}{4} \cdot \frac{3m_J^2 - J(J+1)}{J(2J-1)} \cdot \frac{3m_I^2 - I(I+1)}{I(2I-1)} \, . \tag{12}$$

Here m_I and m_J are the magnetic quantum numbers. The other notation is as in Eq. (3). The $^2P_{3/2}$-state splits up, then, into $(2J+1) \times (2I+1)$ levels. This splitting is depicted in Fig. 9 for the $^2P_{3/2}$-state of K^{39} ($I=3/2$) [48]. The excitation of this state can, for example, be

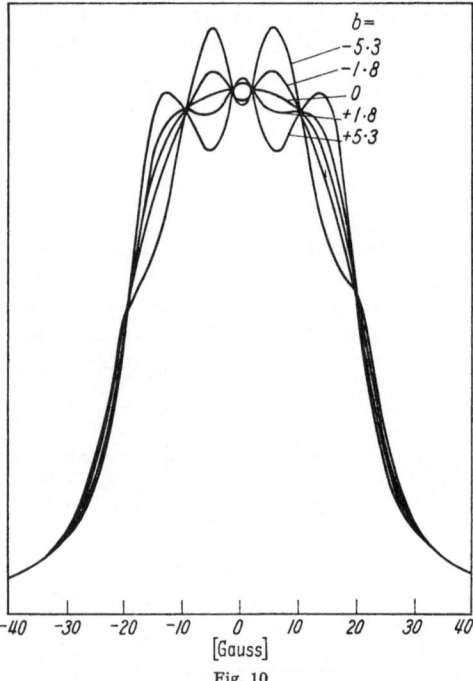

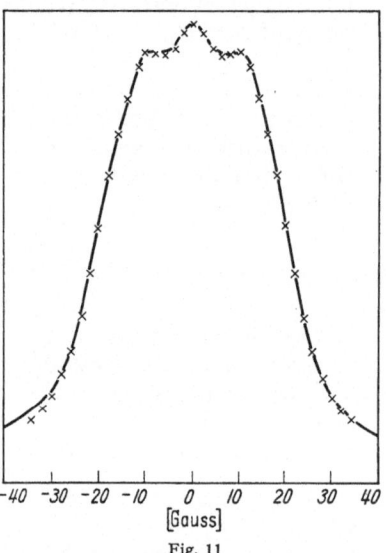

Fig. 10

Fig. 11

Fig. 10. Theoretical resonance curves for the 8 unresolved rf transitions between the Zeeman levels of the hfs in the 3 $^2P_{3/2}$-state of Na^{23} ($I = 3/2$). Excitation with π-light. Depicted is the change of the resonance curve for various values of the quadrupole coupling constant b, which is given in Mc/s

Fig. 11. Theoretical curve for the case depicted in Fig. 10 under the assumption $a = 18.5$ Mc/s, $b = 2.25$ Mc/s. The crosses are the experimentally measured points

carried out with π-light. If the magnitude of the hfs Zeeman splitting is larger than the Doppler width of the exciting spectral line, the intensity of the light source must be spread over a greater spectral region by means of a magnetic field. When rf transitions $\Delta m_J = \pm 1$, $\Delta m_I = 0$ are induced, 8 unresolved resonance signals are observed in the case of Na^{23} and K^{39}. Fig. 10 depicts the superposition curve for the case of Na^{23} (3 $^2P_{3/2}$) calculated under the assumption of various values for b, the quadrupole coupling constant. Fig. 11 shows the agreement of the data points with a fitted theoretical curve [49].

The $^2P_{1/2}$-state of K^{39} was also investigated in a strong magnetic field [50]. Because of the equality of the J-values of the ground and excited states it is only possible to detect signals when circularly polarized light is used for the excitation.

The Breit-Rabi formula serves for the analysis of the experimental data. If the static magnetic field as well as the transition frequency is measured, each of the measured transitions $\Delta m_J = \pm 1$, $\Delta m_I = 0$ can be expressed as a function of $a_{1/2}$ and g_j when the g_I'-term of the formula is neglected. (Such measurements of the static magnetic field are made either with a proton resonance or an optical pumping experiment [31]. The latter method is especially suitable for the precise measurement of small magnetic fields.) If the (approximately) straight lines corresponding to various measurements are plotted on a $a_{1/2} - g_j$ diagram, it is possible to determine $a_{1/2}$ as well as g_j from the point of intersection of lines for various values of m_I. The measurements yielded the value 2/3 (as expected) for the g_j-factor of the $4\,^2P_{1/2}$-state of K^{39}, and $a_{1/2} = 8.99(15)$ Mc/s. As in the case of Rb (see part 1 of this section) the ratio $a_{1/2}/a_{3/2}$ is not in agreement with the theoretical value 5.08 but is instead 4.56(3).

4. Studies of the Sternheimer correction [32—39]

The fine structure of the alkalis is especially suited for the double resonance investigation of a complete series of np-states. In addition there is the advantage of a relatively precise determination of the values of $\langle r^{-3}\rangle_j$ from the fine structure splittings of the ^2P-terms or from the measured a-factor of the $^2P_{3/2}$-state (see part 1 of this section). With this $\langle r^{-3}\rangle_j$ one can then determine the quadrupole moment of the nucleus. Recently the hfs splitting of several excited $^2P_{3/2}$-states of Rb^{85} and Rb^{87} as well as Cs^{133} was measured in order to ascertain a deviation among the Q-values which are independently obtained from the various states. The cause of such a deviation is generally considered to be the Sternheimer effect: the polarization of the electron core by the quadrupole potential of the nucleus. The correction which results from the consideration of this effect exhibits a dependence on the principal quantum number n of the term because the interaction of the valence electron with the polarized electron core takes on different values for various n. The Sternheimer correction has been numerically calculated for various alkalis [36, 39]. The correction factors $(1 - R)^{-1}$ [see Eq. (11)] for Rb are found in Table 1. R is the ratio of the hfs interaction energy

Table 1. *Study of the Sternheimer corrections in Rb*

$^2P_{3/2}$ State	$\dfrac{1}{1-R}$ [36]	Q_{hfs} (Rb^{87}) 10^{-24} cm^2	Q_{corr} (Rb^{87}) 10^{-24} cm^2	Q_{hfs} (Rb^{85}) 10^{-24} cm^2	Q_{corr} (Rb^{85}) 10^{-24} cm^2	Ref.
$5p$	0,787	+0,144(1)	+0,113(1)	+0,298(1)	+0,235(1)	[31, 51]
$6p$	0,827	+0,138(1)	+0,114(1)	+0,286(1)	+0,237(1)	[52][55]
$7p$	0,845	+0,133(3)	+0,112(3)	+0,280(6)	+0,237(6)	[53][54]

of the quadrupole moment which is induced in the core to that of the nuclear quadrupole moment. R can be either positive or negative depending on whether shielding or antishielding of the nuclear quadrupole moment by the electron core predominates.

Also in Table 1 are given the Q-values determined in the various fine structure states along with the corrected (according to STERNHEIMER) values. The correction produces a better agreement. It must be mentioned that the a-factors used for the calculation of the Q-values were not corrected for core polarization. If one uses the fine structure splitting instead of the measured a-factor for this calculation, the magnitude of the Q-values changes ($\approx 5\%$ greater) but the relative differences between the Q's for various values of n remain the same. Furthermore it should be noted that the accuracy of the measurements decreases strongly with increasing principal quantum number. This has its origin above all in the reduction of the amount of resonance light for the transition to the ground state because of (a) competing emissions into other terms as well as (b) the smaller oscillator strength for the absorptive population of the state. The $7\,{}^2P_{3/2}$-state in the Rb-I spectrum decays for example into 6 other states in addition to the ground state (although the individual transitions are not as intense as that to the ground state).

5. Results for nuclear moments in the alkali group

The magnetic moments of the alkalis which have been investigated with double resonance up to now are known more precisely from nuclear resonance experiments. For this reason only Q-values have been determined with double resonance in the alkali group. The a- and b-factors of the excited ${}^2P_{3/2}$-states as well as the calculated (from these factors) nuclear quadrupole moments are summarized in Table 2 (on the following page). The Q-values which are determined in various terms of an atom agree with one another within the error bars except for the Rb values, which agree only after use of the Sternheimer correction. For the Rb and Cs isotopes it was possible, with the exception of the two radioactive atoms Cs^{135} and Cs^{137}, to investigate all three hfs separations of a ${}^2P_{3/2}$-state. In this way a and b were determined by means of three sets of equations*; the agreement between the calculated values was very good. The cases of the Na and K isotopes are much more difficult. Here only the two uppermost hfs separations could be measured or only the two corresponding 2-quantum transitions were observed [58]. However, using the known magnetic moments it is usually possible to calculate the a-factors on the basis of the fine structure splitting well enough (maximum deviation about 10%) to decide questions of level ordering.

---

* Octupole interactions can be neglected for the accuracy of the measurements.

Table 2. *Compilation of hfs measurements for alkali isotopes*

Nucleus	State	Meth.	$a_{Mc/s}$	$b_{Mc/s}$	$Q_{10^{-24} cm^2}$ *	Ref.
Na²³	$3\,^2P_{3/2}$	a**	19,5(6)	2,4(1,4)	+0,10(6)	[56]
		b	18,5(6)	2,25(40)	+0,097(13)	[49]
	$4\,^2P_{3/2}$	a	6,20(12)	1,0(3)	+0,13(4)	[57]
K³⁹	$5\,^2P_{3/2}$	a, b	1,97(10)	1,7(3)	+0,11(2)	[48, 58]
	$5\,^2P_{1/2}$	b	8,99(15)			[50]
K⁴⁰	$5\,^2P_{3/2}$	a	−2,450(46)	−1,31(33)	−0,093(25)	[59]
Rb⁸⁵	$5\,^2P_{3/2}$	a**	25,029(16)	26,032(70)	+0,298(1)	[31, 51]
	$6\,^2P_{3/2}$	a	8,16(6)	8,40(40)	+0,29(2)	[60, 61]
		a	8,178(9)	8,199(40)	+0,286(1)	[52]
		b	8,25(10)	8,16(20)	+0,283(8)	[31]
	$7\,^2P_{3/2}$	a	3,72(3)	3,65(10)	+0,280(6)	[53]
Rb⁸⁷	$5\,^2P_{3/2}$	a**	84,852(30)	12,611(70)	+0,144(1)	[31]
	$6\,^2P_{3/2}$	a	27,63(10)	4,06(20)	+0,14(1)	[60, 61]
		a	27,707(15)	4,000(39)	+0,140(1)	[52]
		a	27,70(2)	3,94(4)	+0,138(1)	[55]
	$7\,^2P_{3/2}$	a	12,58(2)	1,72(4)	+0,133(1)	[53]
Cs¹³³	$7\,^2P_{3/2}$	a	16.60(1)	−0,11(8)	−0,003(2)	[42, 62]
		a	16,609(5)	−0,16(6)	−0,0036(13)	[45, 63]
	$7\,^2P_{1/2}$	a	100,2(3)			[46]
	$8\,^2P_{3/2}$	a	7,626(5)	−0,049(42)	−0,0024(20)	[64]
Cs¹³⁵	$7\,^2P_{3/2}$	a	17,576(6)	2,19(9)	+0,049(2)	[45, 63]
Cs¹³⁷	$7\,^2P_{3/2}$	a	18,280(6)	2,23(9)	+0,050(2)	[45, 63]

a = Zero magnetic field.
b = intermediate or strong magnetic field.
* Without Sternheimer correction.
** These states were also investigated by atomic beam resonance. See Ref. [75—77].

IV. Determination of the quadrupole moments of stable alkaline earth nuclei with odd-A

The ground state $^1S_0$ for the I-spectra of the alkaline earths exhibits a spherically symmetric electronic charge distribution and thus has no interaction with the electric quadrupole moment of the nucleus. For this reason the hfs splittings of the first excited P-states, $^3P_1$ and $^1P_1$, are especially important for the determination of Q with double resonance. As a matter of preference one investigates that state of the sp-configuration for which the ratio of the hfs splitting to the line width of the rf signal is largest. For the $^1P_1$-state one obtains a large line width of the rf signal (because of the short lifetime of the state) along with a small hfs splitting. On the other hand, for the decay of the $^3P_1$-state into the $^1S_0$-ground state the forbiddenness of intercombination lines ($\Delta S = 1$ for pure Russel-Saunders-coupling) tends to slow the transition. This state has therefore a relatively long lifetime and a correspondingly small line width. It also has a large hfs splitting due to the s-electron.

1. Double resonance experiments in excited $^3P_1$-states of the alkaline earths (group 2a of the periodic table) [65—69]

Double resonance studies of the alkaline earths offer a series of problems which are not present for the investigation of the alkalis. First

of all the difficulty in evaporating them must be mentioned. With that
the problem arises of finding a cell which can endure high temperatures,
which is not attacked by the alkaline earth vapors and which possesses
good optical transparency. In addition the wave lengths of the intercombination lines ($^3P_1 - {}^1S_0$) of Ca, Sr and Ba lie in the red or infrared
region so that the spectral component of the heat radiation in this
region produces a troublesome background. The small oscillator strengths
of the intercombination lines make necessary a large vapor density of
the absorber in order to obtain sufficient resonance light. To avoid these
difficulties with the alkaline earths the light source used was usually a
hollow cathode, the absorber an atomic beam. Because of the large
amounts of an isotope needed with such a setup it was not feasible to
use enriched isotopes. The alkaline earth isotopes which are susceptible
to investigation are however of at most only 10% natural abundance and
have a hfs splitting which is considerably larger than the line width of
the exciting spectral line. Consequently the light of the even-A isotopes,
which have the greatest abundance in the light source, cannot contribute
to the population of hfs states via resonance absorption if special
measures are not undertaken.

a) Excitation relations for the lowest $^3P_1$-state of the alkaline earths

In the following the problem of the Ba-I spectrum will be discussed
in some detail as an example. For the excitation of the $6\,^3P_1$-state the
spectral line $\lambda = 7911$ Å ($6\,^1S_0 - 6\,^3P_1$), produced in a hollow cathode,
was available. An atomic beam served as the absorber (see Fig. 12 and 14).
The light source and absorber contained barium of natural composition
[82.1% even Ba isotopes, 6.6% Ba^{135} ($I = 3/2$) and 11.3% Ba^{137} ($I = 3/2$)].
With the light source used the Doppler width of the exciting spectral line
amounts to about 600 Mc/s while the separations α_F of the three hfs
levels $F = 5/2$, $3/2$ and $1/2$ from the center of gravity of the state are
about 2—5 times so large (see Fig. 12). Under these conditions one must
reckon with the fact that the uneven Ba isotopes in the atomic beam
will absorb only that light which the uneven isotopes of the light source
emit. (The difference in hfs splitting for the two uneven Ba isotopes is
small compared with the Doppler width and can be considered 0 for
this discussion; the isotope shift is also small compared with the Doppler
width.) If 1 represents the resonance light which is reemitted from the
even isotopes in the atomic beam, then for the two uneven isotopes the
sum of the resonance light which is radiated from the level F is

$$\beta = V^2 \cdot \left(\frac{2F + 1}{\sum\limits_F (2F + 1)} \right)^2, \tag{13}$$

where $V = 0.218$ is the ratio of the abundance of the odd (17.9%) to
that of the even (82.1%) isotopes. For the $F = 3/2$ hfs level of barium
$\beta \approx 5 \times 10^{-3}$. In order to obtain more resonance light from the odd
isotopes the spectral line $\lambda = 7911$ Å ($^1S_0 - {}^3P_1$) produced by the light
source was subjected to a Zeeman splitting of such a magnitude that one

of the Zeeman levels of the fine structure of the even isotopes coincided with a hfs level of the odd isotopes. Fig. 12 shows this state of affairs. At the left the Zeeman splitting of the even isotopes in the light source is depicted. This splitting corresponded to the separation of the $F = 3/2$ hfs level from the center of gravity of the state. The right side of the figure shows the hfs. The direction of the magnetic field H_L at the light source is chosen parallel to that of the light emitted from the hollow cathode so that only the two σ-components of the spectral line of the even isotopes are emitted. Then, under the assumption that the total intensity is not changed by the Zeeman splitting, the gain in the excitation of a single hfs level is $G = 1/(2\sqrt{\beta})$. A similar technique was first applied

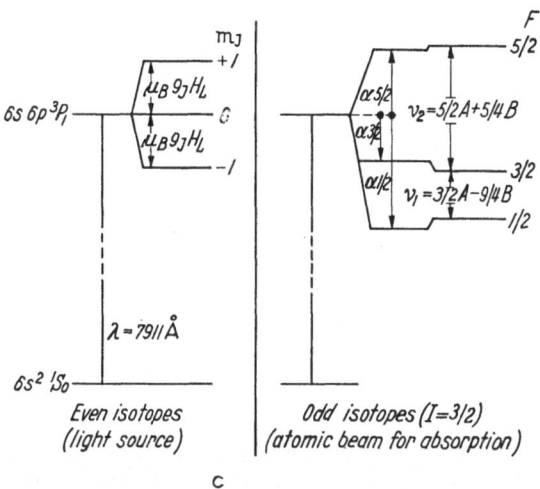

Fig. 12. Term scheme of the $6s\,6p\,{}^3P_1$-state in Ba I. At the left the Zeeman splitting for the even isotopes in the light source which is made so large that one of the Zeeman levels coincides with a hfs level of the odd isotopes in the atomic beam. At the right the term scheme for the odd Ba isotopes ($I = 3/2$)

to the ${}^3P_1$-state of Hg[73]. It possesses the advantage that one obtains an enhanced excitation of the single hfs level simultaneously with the disappearance of the resonance light from the even isotopes, the latter having been a source of noise. The experimental test of this improvement in the excitation is given by the observation of rf transitions $\Delta F = 0, \Delta m_F = \pm 1$ which can be resolved from one another because of the difference in the g_F-factors of various F-levels. Fig. 13 is a plot of the intensity of the rf transitions $\Delta F = 0, \Delta m_F = \pm 1$ for two of the hfs levels as a function of the Zeeman

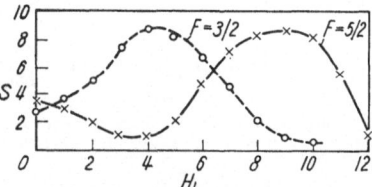

Fig. 13. Signal S for rf transitions $\Delta F = 0$, $\Delta m_F = \pm 1$ in a hfs Zeeman multiplet as a function of the magnetic field H_L which produces the Zeeman splitting in the light source. S is a measure for the excitation of a hfs level

splitting of the light source. This plot depicts, then, the population of a hfs level. In the case of Ba one can achieve a maximum population of the one hfs level with a minimum population of the two other levels and thus a truly marked gain in the excitation of the hfs states [68].

b) Experimental setup for the investigation of alkaline earths

The experimental arrangement shown in Fig. 14 served for the measurement of the hfs splitting of the lowest excited ${}^3P_1$-states of Sr⁸⁷ as well as Ba[135] and Ba[137]. This setup can be applied with advantage to all cases where the substance is either not easily evaporable or reacts with

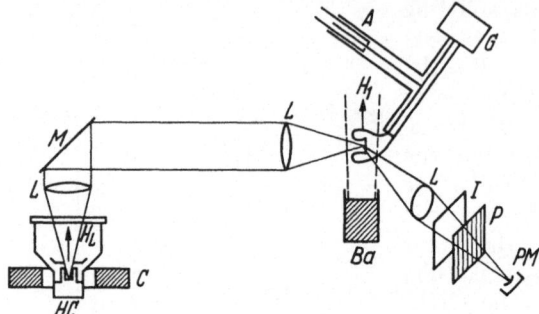

Fig. 14. Atomic beam experimental arrangement. *HC*: hollow cathode light source, *C*: coil which produces the magnetic field H_L, *L*: lenses, *M*: mirror, H_1: direction of the rf-magnetic field, *A*: adjustable stub, *G*: frequency generator, *I*: interference filter, *P*: polarizer, *PM*: photomultiplier

the wall material of the cell to quickly reduce the optical transparency. The light source was a hollow cathode which contained the alkaline earth in metallic form. The Zeeman splitting of the spectral line emitted by the hollow cathode required magnetic fields up to 1500 Gauss which were produced in a flat coil *C*. In all of the hfs measurements the magnetic field was adjusted so that the middle hfs level of the uneven isotope could be maximally populated. In this way the strongest $\Delta F = \pm 1$ rf signal was obtained. An intense atomic beam of Sr or Ba served for the resonance absorption of the exciting light. The resonance light was observed

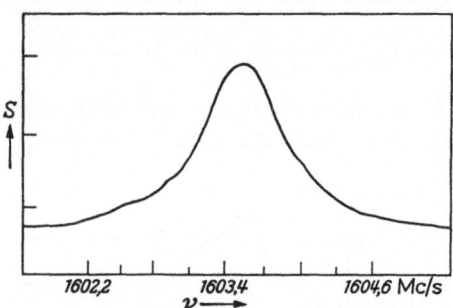

Fig. 15. Registered line shape of the rf transition $F = 3/2 \leftrightarrow F = 1/2$, $6s\,6p\,{}^3P_1$ for Ba[135]

perpendicular to the direction of the incoming light with a red-sensitive photomultiplier. The rf-magnetic field H_1 for the induction of hfs transitions was arranged perpendicular to both the direction of the incoming light and that of the observed light. The rf-field with frequencies of 1000–3000 Mc/s was produced by a signal generator and fed into a coil located in the resonance region by means of a coaxial conductor. The magnetic field of the earth and stray magnetic fields had to be compensated out very painstakingly in order to avoid a shift of the resonance frequency [68]. This shift results from the asymmetric broadening of the rf signal in a magnetic field when excitation of a single hfs term with σ^+- or σ^--light is used.

Fig. 15 shows as an example the registered curve of the rf transition from $F = 3/2$ to $F = 1/2$ for Ba[135]. The same transition was resolved into its individual components in a weak magnetic field. This process is well

suited for the identification of the signal observed in zero-field. Fig. 16
shows a corresponding experimental curve for Ba^{137}. For the σ^+-excitation

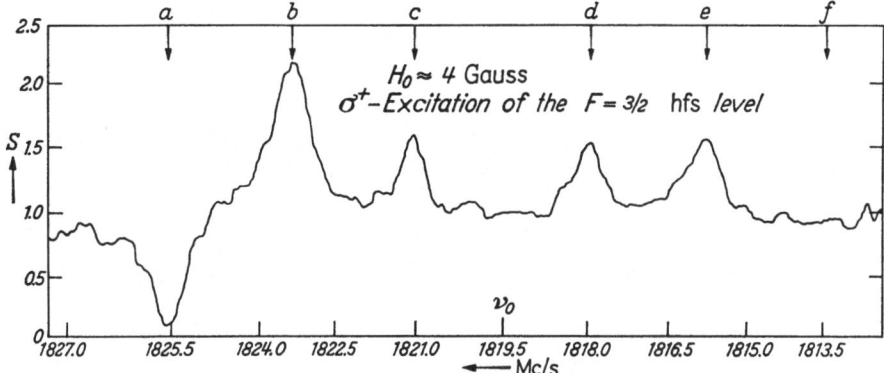

Fig. 16. Rf transition $\varDelta F = \pm 1$, $\varDelta m_F = 0$, ± 1 from the $F = 3/2$ to the $F = 1/2$ hfs level of the $6\,^3P_1$-state
of Ba^{137} in the presence of an external magnetic field $H_0 \approx 4$ Gauss. ν_0 denotes the position of the resonance
for $H_0 = 0$. Positions of the rf signals:

	from $F = 3/2$	to	$F = 1/2$
a)	$m_F = +3/2$		$m_F = +1/2$
b)	$m_F = +1/2$		$m_F = +1/2$
c)	$m_F = -1/2$		$m_F = +1/2$
d)	$m_F = +1/2$		$m_F = -1/2$
e)	$m_F = -1/2$		$m_F = -1/2$
f)	$m_F = -3/2$		$m_F = -1/2$

of the $F = 3/2$ level without excitation of the $F = 1/2$ level the $F = 3/2$,
$m_F = -3/2$ Zeeman state is not populated. The corresponding rf
transition f is not seen in Fig. 16.

2. Calculation of the nuclear quadrupole moments from the hfs of the $^3P_1$-state [6, 70]

In the calculation of Q from the B-factor of a $^3P_1$-state one must bear
in mind that the state $^1P_1$ (which has the same J) also appears for the
electronic configuration. This $^1P_1$-state can be mixed in with the $^3P_1$-state.
In the calculation of this mixing one starts with eigenfunctions in either
pure Russel-Saunders or pure jj coupling. If one uses the latter, the eigenfunction of the state labelled $^3P_1$ has the form

$$\psi = c_1 \psi_1 (j = 1/2; j = 3/2)_{J=1} + c_2 \psi_2 (j = 1/2; j = 1/2)_{J=1} \qquad (14)$$

where ψ_1, ψ_2 signify the eigenfunctions with $J = 1$ for the case of pure jj
coupling. The coefficients c_1 and c_2 can be obtained in the following ways:

a) from the position of the nsnp P-states in the level scheme [80]

In intermediate coupling the coefficients c_1 and c_2 may be written as

$$c_1 = \sin(\vartheta_0 - \vartheta) , \quad c_2 = \cos(\vartheta_0 - \vartheta) \qquad (15)$$

with $\vartheta_0 = \text{arc tg} \sqrt{l/(l+1)}$ and $\sin^2\vartheta = \varDelta/D$. Here l is the orbital angular momentum quantum number, D the $^1P_1 - {}^3P_1$ separation, and $\varDelta$ the deviation of the $^3P_1$-level from its position in pure Russel-Saunders coupling. For pure Russel-Saunders coupling one may write

$$c_1 = \sqrt{l/(2l+1)} \quad \text{and} \quad c_2 = \sqrt{(l+1)/(2l+1)} \ . \tag{16}$$

In using the c_i-factors for the calculation of the electric field gradient it is necessary to bear in mind that in addition to the spin-orbit interaction the spin-other orbit and spin-spin interactions also occur.

b) from the lifetime of the states $^3P_1$ and $^1P_1$ [72]

Under the assumption that the $^3P_1$-state is strictly metastable in pure Russel-Saunders coupling one can write

$$\frac{\beta^2}{\alpha^2} = \frac{\tau(^1P_1)}{\tau(^3P_1)} \cdot \frac{\lambda^3(^3P_1 - {}^1S_0)}{\lambda^3(^1P_1 - {}^1S_0)} \tag{17}$$

where τ is the lifetime of the state and $\lambda(^{2S+1}P_1 - {}^1S_0)$ is the wavelength for the transition to the ground state. α and β are the coupling coefficients of the Russel-Saunders eigenfunctions; they are related to c_1 and c_2 through

$$\alpha = c_1 \sqrt{\frac{l}{2l+1}} + c_2 \sqrt{\frac{l+1}{2l+1}}$$

$$\beta = c_1 \sqrt{\frac{l+1}{2l+1}} - c_2 \sqrt{\frac{l}{2l+1}} \tag{18}$$

with the normalisation $\alpha^2 + \beta^2 = 1$.

c) from the g_J-factor of the state

The Landé g_J-factor is changed by the mixing of the states $^1P_1$ and $^3P_1$. The new expression is

$$g_J(^3P_1) = \alpha^2 g_J^{RS}(^3P_1) + \beta^2 g_J^{RS}(^1P_1) + \varDelta g \ . \tag{19}$$

g_J^{RS} are the g_J-factors obtained in pure Russel-Saunders coupling when the quantum electrodynamically correct g-factor of the electron is used; $\varDelta g$ takes into account corrections for the relativistic mass increase of the electron, the spin-orbit coupling, diamagnetic effects and the Hughes-Eckart reduced mass [81].

In the case of the alkaline earths, $\varDelta g$ has the same order of magnitude as the correction for the mixing. Thus it is not possible to make an exact determination of the c's with method c. The c's ascertained with methods a and b can be compared in the case of Zn [71, 78]. The agreement is good. These two methods will then be given priority.

For the calculation of Q from B one must know the value of $\langle r^{-3} \rangle_J$ in addition to the c's (which are ascertained according to one of the processes decribed above). If one takes the value given by the fine structure splitting one obtains the following expression for the second derivative of the electric potential which the p-electron of the $nsnp \ ^3P_1$-

state produces at the nucleus [6, 70]

$$\langle \varphi_{zz}^{(0)} \rangle \, (^3P_1) = \frac{e\,[c_1^2\,R_r'(Z_i) - 2\sqrt{2}\,c_1 c_2 S_r(Z_i)]\,\widetilde{\delta W}}{15\,\mu_B^2\,Z_i\,H_r(Z_i)} \tag{20}$$

with the notation of Eqs. (5), (6) and (7), and where $\widetilde{\delta W}$ = fine structure splitting $(^3P_0 - {}^3P_2)$ and S_r = relativistic correction factor.

As one sees, the exact calculation of the field gradient requires the knowledge of the effective nuclear charge as well as of the value of the fine structure splitting. Both values are not known very precisely — in the case of the fine structure splitting a correction for the spin-spin and spin-other orbit interactions may have to be made [71].

Another possibility for the determination of the value of $\langle r^{-3} \rangle_J$ is available if the A-factors of two or more fine structure states in the same fine structure multiplet are known. This is above all the case for the ^3P-Terms of the elements of group 2b. The $^3P_2$-states of these elements have a higher excitation energy than those of the alkaline earths and can therefore be investigated with the atomic beam techniques which have been developed for metastable states [82, 84]. Recently the A-factor of the $6s\,6p\,{}^1P_1$-state of Ba was accurately measured with optical spectroscopy [85]. The relevant A-factors can be expressed in terms of the a-factors of the electrons in the configuration (in this case a_s and a_j) as follows [6, 70, 86].

$$A\,(^3P_2) = \frac{1}{4}\,a_s + \frac{3}{4}\,a_{3/2}, \tag{21}$$

$$A\,(^3P_1) = \frac{1}{4}\,(2c_2^2 - c_1^2)\,a_s + \\ + \left[\frac{5}{4}\,c_1^2 - \frac{5}{16}\,\sqrt{2}\,c_1 c_2 \xi + \frac{5}{2}\,\Theta\,c_2^2\right]\,a_{3/2}. \tag{22}$$

The A-factor of the $^1P_1$-state is found by making the replacement

$$c_1' = c_2, \quad c_2' = -c_1 \tag{23}$$

in Eq. (22), c_1' and c_2' being the coupling coefficients for this state. ξ and Θ are relativistic correction factors which SCHWARTZ [80] defines. The A-factor of the $p_{3/2}$-electron is [Eq. (5b)]

$$a_{3/2} = \frac{16}{15}\,\frac{g_I'}{h}\,\mu_0^2\,F_{r,s}\,\langle r^{-3} \rangle_j. \tag{24}$$

One can then use the measurements of the A-factors of two $nsnp$ P-states to eliminate a_s and thus to find $a_{3/2}$ and $\langle r^{-3} \rangle_j$. It must be remembered that the measured A-factors may perhaps need correcting for core polarization [30] and for screening by the electron core [34]. Finally an example of each of the two methods. In the case of the BaI-spectrum Q^{137} is calculated, using the A-factors of the $^3P_1$ and $^1P_1$ states, to be $Q\,(^3P_1) = +0.31 \cdot 10^{-24}$ cm^2 from the hfs of the $6s\,6p\,{}^3P_1$-state. Using the fine structure splitting with the corrections given by H. WOLFE one obtains $Q\,(^3P_1) = +0.30 \cdot 10^{-24}$ cm^2 [68]. This very nice agreement is however probably accidental: the Q-values calculated with the A-factor method differ by 30% in the two states $6\,^1P_1$ and $6\,^3P_1 (Q\,(^1P_1) = 0.21 \times$ $\times 10^{-24}\,\text{cm}^2)$. Another check of the two methods for determining $\langle r^{-3}\rangle_J$ is given by Cd, where the 2 values of Q obtained from the $^3P_1$-state differ by 25%. Nevertheless it appears desirable to know all the A-factors and hfs splittings of a configuration when calculating Q.

3. Results of double resonance studies of the alkaline earths

As of the writing of this article the hfs splitting of the stable alkaline earths Sr[87], Ba[135] and Ba[137] has been measured. The measurements yielded exact values of the A- und B-factors because of the small line width of the rf signal (some 100 kc/s) coupled with the large hfs splitting in the $^3P_1$-state (1000–3000 Mc/s). In the case of Ba it was also possible to determine the hfs anomaly $\Delta\varepsilon$ [6, 80, 87] since the magnetic moments of the nuclei [74] are known. $\Delta\varepsilon$ is defined as follows:

$$\Delta\varepsilon = \left(\frac{A'}{A''} \cdot \frac{I'}{I''} \cdot \frac{\mu_I''}{\mu_I}\right) - 1 \,, \tag{25}$$

where the superscript primes are used to denote two isotopes of the element. The A's and B's listed in Table 3 were calculated directly from the measured hfs separations without making corrections for second order effects. (See Ref. [72] for a summary of these effects.)

Table 3. *Compilation of hfs measurements for alkaline earth isotopes* (group 2a)

Isotope	State	Method	$A_{\text{Mc/s}}$ [*]	$B_{\text{Mc/s}}$ [*]	$Q_{10^{-24}\,\text{cm}^2}$ [**]	Ref.
Sr[87]	$5s5p\,^3P_1$	Zero field measurements	—260,084(2)	—35,658(6)	+0,36(3)	[67, 69]
Ba[135]	$6s6p\,^3P_1$	Zero field measurements	1028,31(2)	—27,08(2)	+0,18(2)	[66, 68]
Ba[137]	$6s6p\,^3P_1$	Zero field measurements	1150,59(2)	—41,61(2)	+0,28(3)	[66, 68]
	$6s6p\,^1P_1$	Level crossing	—113,2(1,0)			[131]

   [*] Uncorrected for second order effects.
 [**] Without Sternheimer correction.

V. The elements of group 2b

The atoms zinc, cadmium and, above all, mercury were from the beginning the preferred objects for double resonance experiments. This is due to the fact that their atomic spectra are especially suited for such investigations. And, in addition, the metals themselves are easier to handle than the alkalis and the alkaline earths. Furthermore it was possible to produce many of the radioactive nuclei of the 2b-elements with half lives of more than a few hours through (p, n) reactions in the stable copper, silver or gold nuclei – and to produce them in sufficient quantities so that their nuclear moments could be determined. Also, in the case of the stable nuclei of the 2b elements, atomic beam studies of the $^3P_2$-state of the same configuration have often been undertaken; this increases the certainty of the calculation of the moments. The range of applicability was considerably increased by the level crossing method which brought additional hfs data. In this section double resonance experiments on the elements of group 2b will be discussed. The next section will be devoted to the level crossing experiments.

The determination of the hfs splitting of the $^3P_1$-states required special measures particularly in the case of mercury. This was so because the 6s-electron of the Hg I-spectrum produces a very large hfs splitting (it can be resolved optically [88, 89]). And this along with a relatively large deviation from RS coupling and thus with a short lifetime for the $^3P_1$-state. For the direct measurement of the hfs splitting in zero-magnetic field it is necessary to have a signal generator which can produce rf-magnetic fields of some gauss at frequencies up to 15000 Mc/s. In order to avoid this difficulty the Zeeman hfs transitions were often investigated with I and J partially decoupled. For the analysis of such measurements the following expressions are useful. The Hamiltonian for an atom in an external magnetic field may be written

$$\mathscr{H} = A\mathbf{I}\mathbf{J} + B\frac{3(\mathbf{I}\mathbf{J})^2 + (3/2)\,\mathbf{I}\cdot\mathbf{J} - \mathbf{I}^2\mathbf{J}^2}{2I(2I-1)\cdot J(2J-1)} +$$
$$+ \mu_0\,g_J\,\mathbf{J}\cdot\mathbf{H}_0 + \mu_0\,g_I'\,\mathbf{I}\cdot\mathbf{H}_0\,. \tag{26}$$

(W_J, the fine structure energy, is omitted in this equation.)

For the case $H_0 = 0$ this reduces to Eq. (3), for the decoupling of I and J in a strong magnetic field this gives Eq. (12). If the Zeeman splitting within an F-level is small compared to the hfs splitting, the additional terms in the Hamiltonian can be treated as a perturbation. The energy of a hfs Zeeman sublevel is — as long as I and J are well-coupled to F —

$$W_{F,m_F} = W_F + \mu_0\cdot H_0\cdot m_F\cdot g_F \tag{27}$$

with W_F as in Eq. (3) and

$$g_F = g_J\frac{F(F+1)+J(J+1)-I(I+1)}{2F(F+1)} +$$
$$+ g_I'\frac{F(F+1)+I(I+1)-J(J+1)}{2F(F+1)}\,. \tag{28}$$

For this weak-field case the Zeeman energy differences do not depend on Q. With the partial decoupling of I and J in intermediate magnetic fields the Zeeman sublevels deviate from this linear dependence. For the calculation of the energy level diagram the eigenvalue problem of the Hamiltonian must be solved exactly. The case $I = 1/2$ or $J = 1/2$ leads to the well-known Breit-Rabi formula; when $I, J > 1/2$ the solution of the general secular equation is necessary in order to extract A and B from the data [6]. Furthermore the data are still subject to second order corrections (which are compiled in Ref. [72] for the ^3P-terms which are of interest here).

1. Scanning experiments with Hg isotopes

The intercombination line $\lambda = 2537$ Å of the Hg I spectrum exhibits for the various isotopes of Hg an isotope shift which is larger than the Doppler width of the spectral line. However, since — in addition to the even Hg isotopes 196, 198, 200, 202 and 204 — the two odd isotopes Hg[199] $(I = 1/2)$ and Hg[201] $(I = 3/2)$ are contained in naturally-occurring

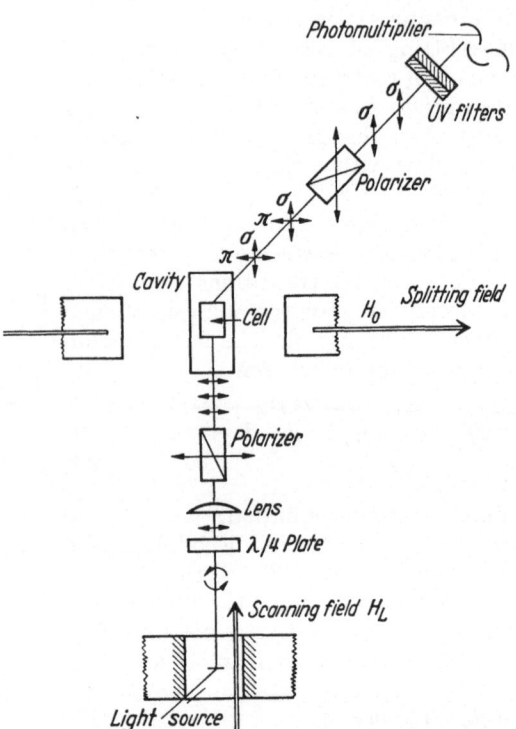

Fig. 17. Schematic of setup for a scanning experiment

mercury and since their hfs is large compared with the Doppler width, one obtains an unresolved structure for the line when one uses spectroscopic methods without separated isotopes. SAGALYN, MELISSINOS and BITTER [73] employed a combination of the double resonance method with a magneto-optical scanning method in order to resolve this overlapping structure. This procedure will be illustrated with the example of the even isotopes.

The light source contains the even isotope Hg[198] in a state of high enrichment (see Fig. 17). The emitted intercombination line $\lambda = 2537$ Å is split into a Zeemann multiplet by applying a magnetic field parallel to the direction of emission. With this arrangement only the two σ-components of the spectral line are observed. One of the σ-components is linearly polarized with a quarter-wave plate and absorbed with an additional polarizer; the other is linearly polarized and transmitted. One has thus a light source in which the wave length of the emitted spectral line can be varied with a magnetic field. This light enters a resonance cell which contains the even and odd isotopes in a static magnetic field. The excitation conditions are chosen so that in each case only the sublevel with $m_J = 0$ (whose position is field-independent) is populated. Rf transitions are induced in a resonance cavity which contains the absorption cell. If one induces the transitions $\Delta m_J = \pm 1$, the signal intensity is proportional to the population of the Zeeman sublevel $m_J = 0$. This follows because the π-excitation and the size of the Zeeman splitting (in this case 3000 Mc/s) do not allow the population of the $m_J = \pm 1$ sublevels. Also, one receives no signals from the odd isotopes since their hfs Zeeman transitions have a different resonance frequency.

Thus one varies the magnetic scanning field and records the intensity of the pertinent resonance signal of the even isotopes. When $g_J(^3P_1)$ and the magnetic scanning field are known, one can in this way determine the relative (to Hg^{198}) positions of the fine structure terms of the even isotopes for sufficiently large isotope shift. Fig. 18 is an example of a

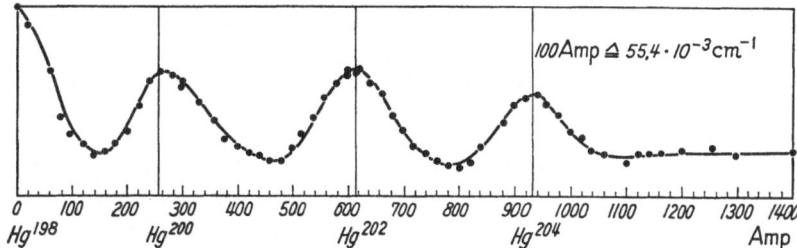

Fig. 18. Scanning curve for even Hg isotopes. The light source contains Hg^{198}

scanning curve for the even Hg isotopes. The width of the observed signals corresponds to the Doppler width. The great advantage of this method is that the uneven Hg isotopes make no contribution to the signal in spite of their presence in the resonance cell.

It is also possible to ascertain the position of the hfs terms by the induction of hfs Zeeman transitions. The sublevels $m_F = \pm 1/2$ of the $F = 3/2$ level of Hg^{199} can be populated with π-excitation and the transitions to $m_F = \pm 3/2$ can be observed. If one measures in addition the transition frequencies in an intermediate magnetic field in a double resonance experiment (which yields more exact values than a scanning experiment does), the position of the hfs level relative to the fine structure term of Hg^{198} is then determined. The Hg^{201} isotope was investigated similarly. Moreover one can use the scanning technique to determine the hfs level ordering and so the sign of the magnetic moment of an isotope. Besides the stable Hg isotopes the radioactive atoms Hg^{197} (65 h) [90], Hg^{197m} (24 h) [91], [133] and Hg^{193} (6h) [92] have also been investigated. Table 4 shows the results for the relative positions of the isotopes (see also Ref. [93]).

Table 4. *Isotope shift relative to Hg^{198} of the line $\lambda = 2537$ Å of the HgI spectrum, measured with the scanning method* [1 mK = 10^{-3} cm^{-1}]

Isotope	Isotope shift [mK]
193	+ 350(26)
196	+ 137(4)
197	+ 91(5)
197m	+ 75(5)
198	0
199	− 16(4)
200	−156(3)
201	−216(3)
202	−339(3)
204	−519(3)

2. Detection of double resonance by frequency change [14]

For the detection of rf transitions $\Delta F = \pm 1$ between the hfs levels of the $^3P_1$-state of Hg^{201} ($I = 3/2$) one can use the frequency change of

the resonance light. Fig. 19 illustrates the method. The hfs splitting
of the $^3P_1$-state is large compared with the Doppler width of the spectral

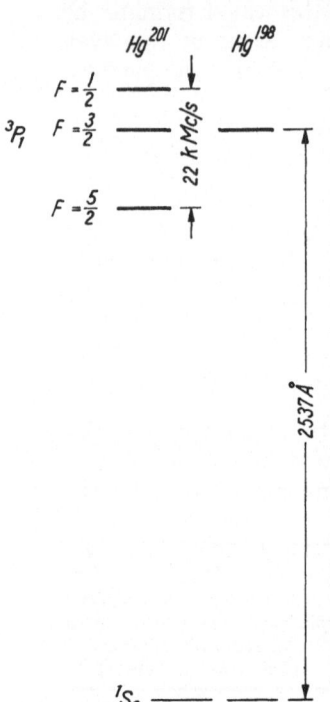

line $\lambda = 2537$ Å. The $F = 3/2$ level lies close to
the $^3P_1$-state of the even isotope Hg^{198}. Therefore only the $F = 3/2$ level is populated by excitation with a light source which contains only
the Hg^{198} isotope. If one absorbs the reemitted
resonance light in a cell which likewise contains
Hg^{198}, then only a fraction of the resonance light
is detected. With the induction of rf transitions
to the $F = 5/2$ or $F = 1/2$ level, the resonance
light contains a component whose wave length
corresponds to the transition from these levels
to the ground state. This light traverses the absorption cell in front of the photomultiplier without significant loss in intensity. Thus the rf
transition is detected through the increase in
the light reaching the photomultiplier. This
method is extremely sensitive for the detection
of rf transitions since (a) all the light contributes
to the signal (not just that with a definite polarization) and (b) practically all detected quanta
belong to the signal, no background contributing
to the noise. Fig. 20 shows the setup for the
Hg^{201} experiment. A similar investigation [94]
was used later for Hg^{197} and Hg^{199}. The production of the rf-field strengths at frequencies
up to 14000 Mc/s was accomplished in a
resonance cavity which contained the resonance
cell. The cavity was set at a fixed frequency

Fig. 19. The excitation of the F
$= 3/2$ hfs level of the $^3P_1$-state of
Hg^{201} with radiation which is
mitted from a Hg^{198} light source.

near the resonance frequency for $H_0 = 0$. A weak homogeneous
magnetic field shifted the Zeeman sublevels until the resonance condition

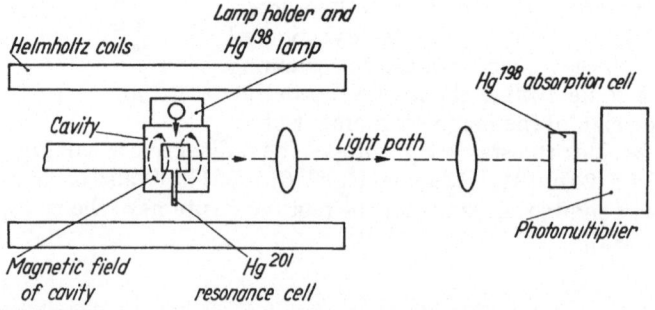

Fig. 20. Experimental setup for the detection of double resonance signals by means of the frequency change
of the resonance light

was fulfilled. The hfs splitting was extrapolated from the measured
frequencies and magnetic fields. The accuracy obtained was very high,
about $1 \cdot 10^{-6}$.

3. Double resonance with stable and radioactive Zn, Cd and Hg isotopes

The hfs of the stable Zn, Cd and Hg isotopes was investigated in the first double resonance experiments. Since mercury exhibits very large hfs splittings for $6\,^3P_1$, an investigation of its isotopes was usually carried out in intermediate magnetic fields where the problems of rf techniques are not so difficult [95, 73, 90, 91]. For the cadmium isotopes Cd^{111} and Cd^{113} (both $I = 1/2$) exact values for the A-factors of the $5\,^3P_1$-state were obtained in a double resonance experiment. The knowledge of these A-factors is of special interest when the magnetic moments of the nuclei are also known. These A-values can then be used for the determination of the magnetic moments of radioactive nuclei whose hfs has been measured. The hfs of the $4\,^3P_1$-state of Zn^{67} was measured in zero external magnetic field [97]. In this case the splitting for the $^3P_1$-state was smaller than in Hg and Cd. The higher degree of RS coupling for Zn leads to a smaller line width and so requires smaller rf-field strengths for the induction of rf transitions.

The extension of the experiments to radioactive nuclei by BITTER and his co-workers for Hg and by NOVICK and his co-workers for Cd resulted in values for previously unknown nuclear spins, magnetic moments and quadrupole moments. With (p, n) or (d, 2n) reactions in stable gold and silver isotopes as well as (n, γ) reactions in stable Cd isotopes and with subsequent mass separation it was possible in many cases to produce sufficient quantities of radioactive isotopes. In doing this it was not so easy to attain high isotopic purity because Hg and Cd are relatively abundant elements whose stable isotopes can easily "contaminate" the resonance cells. The numbers of atoms required for the experiments were extremely small. In general 10^{11} atoms were sufficient*. A description of how such cells of radioactive isotopes are made is found in Ref. [99].

The investigation of the $5\,^3P_1$-state of Cd^{109} (470 d) $(I = 5/2)$ will serve as an illustration of a double resonance experiment for the determination of the nuclear spin and of the sign and magnitude of the magnetic dipole as well as the electric quadrupole moment. The isotope

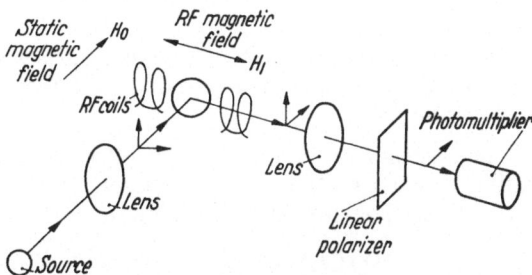

Fig. 21. Experimental setup for the investigation of stable and radioactive Cd isotopes

was made from silver by means of a (p, n) reaction. It was brought into the cell by means of vacuum distillation. The course of the distillation was controlled by observing the activity with a proportional counter. The identification of the radioactive isotope as Cd resulted from the observation of the resonance radiation from the cell. This method is a

---

* In atomic beam resonance experiments even fewer radioactive atoms are required because the radioactive decay can be used for the detection [98].

very sensitive indicator for the presence of the element in the resonance volume.

Fig. 21 is a schematic of the experimental setup. The light source was an rf-lamp for which the Cd intercombination line $\lambda = 3261$ Å was very intense. The source contained Cd of natural composition since some of the Cd^{109} hfs levels could be satisfactorily populated so. A static magnetic field H_0 directed parallel to the incoming light was present over the resonance volume. The magnitude of H_0 was measured with proton resonance. The orientation of H_0 assured that the excitation of Cd atoms took place with σ-light. In order to achieve a sufficiently high vapor pressure the resonance cell was heated by a gas flame in an oven. This is preferable — when no very high degree of temperature stability is required — because of the absence of electronic heating with its accompanying stray fields. The resonance scattering was observed by means of a photomultiplier. A polarizer was placed between the cell and photomultiplier so that only π-light was detected. H_1, the rf-magnetic field, was parallel to the direction of observation of the resonance light.

a) Spin determination in weak magnetic field

The hfs Zeeman resonances in a weak field served for the determination of the nuclear spin of the Cd isotope. The g_F-factors for the three hfs levels are only dependent on g_J, g_I' and I. The contribution of g_I' is so small that it can be neglected. If one measures (by means of hfs Zeeman transitions) two or all three g_F-factors, one obtains a unique value for the nuclear spin from the ratio of these numbers. A resonance in only one F-level does not yield a unique result since the same resonance condition appears in another F-level for another spin. For Cd^{109} there was obtained in this way $g_F/g_J = 0.28$, 0.11 and 0.40. This fixes the nuclear spin as $I = 5/2$.

b) Determination of the spin, A- and B-factors and the sign of the magnetic moment through measurements of hfs Zeeman transitions in an intermediate field

The second order hfs Zeeman effect in intermediate magnetic field splits up each of the resonance signals which were seen in weak field into $2F$ components with the same separation. The separations

$$\Delta \nu (F) = (\nu_{F, m_F+1} - \nu_{F, m_F}) - (\nu_{F, m_F} - \nu_{F, m_F-1})$$

are given in frequency units by

$$\Delta \nu (I+1) = \frac{-2I}{(2I+1)(I+1)^2} (\gamma H_0)^2 \frac{1}{\Delta_1}, \tag{29}$$

$$\Delta \nu (I) = \frac{2I}{(2I+1)(I+1)^2} (\gamma H_0)^2 \frac{1}{\Delta_1} - \\ - \frac{2(I+1)}{(2I+1)I^2} (\gamma H_0)^2 \frac{1}{\Delta_2}, \tag{30}$$

$$\Delta \nu (I-1) = \frac{2(I+1)}{(2I+1)I^2} (\gamma H_0)^2 \frac{1}{\Delta_2} \tag{31}$$

where γ is the gyromagnetic ratio and $\varDelta_1$, $\varDelta_2$ are the hfs separations in external field $H_0 = 0$ which are given by Eq. (3).

From the number of resolved resonances in the hfs Zeeman levels it is possible to determine the F-values and thus (when J is known) I. The separations between the resonances allow the calculation of $\varDelta_1$ and $\varDelta_2$ when the g_J-factor is known and the magnetic field H_0 measured. From $\varDelta_1$ and $\varDelta_2$ are obtained, in the normal way, A and B. If A and μ_I

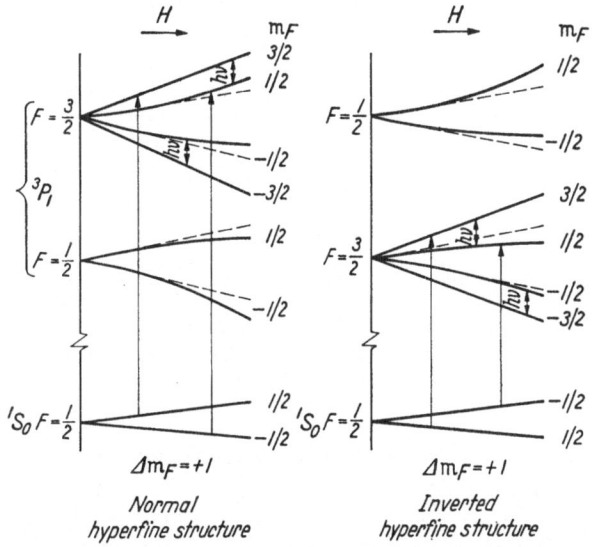

Fig. 22. Determination of the ordering of the hyperfine levels with σ^+ circularly polarized exciting light. In the case illustrated ($I = 1/2, F = 3/2$), this illumination results in the elimination of the low-field resonance if the hfs is normal and elimination of the high-field resonance if the levels are inverted

for another isotope of the element are known, the moment μ_I of the investigated nucleus can be calculated from the measured A-factor with an accuracy limited by the hfs anomaly and corrections.

The question of the sign of μ_I, i. e., of the character — normal or inverted — of the term scheme still remains. The answer can be given with optical methods. As an illustration let us take the case of Cd^{113}. It is especially easy to consider because $I = 1/2$. Fig. 22 depicts the hfs splitting of the $^3P_1$ state of Cd when $I = 1/2$ for the case of normal hfs (left) and inverted hfs (right). If circularly polarized light σ^+ is used to excite the $F = 3/2$ level only the sublevels $m_F = +3/2$ and $m_F = +1/2$ will be populated. Only transitions between these two sublevels could be detected since the transition $m_F = +1/2$ to $m_F = -1/2$ could not be observed with the experimental setup. Because of the mutual repulsion of sublevels which have the same m_F the frequency for the transition $(F = 3/2, m_F = +3/2)$ to $(F = 3/2, m_F = +1/2)$ will be smaller than the corresponding linear Zeeman effect frequency if the hfs is normal. If the hfs is inverted this inequality is reversed. Consequently, in order to determine the level ordering and thus the sign of A, it is only necessary to observe which resonance disappears when σ^+-excitation is used. Fig. 23 shows such results for Cd^{113}. Above are seen with unpolarized

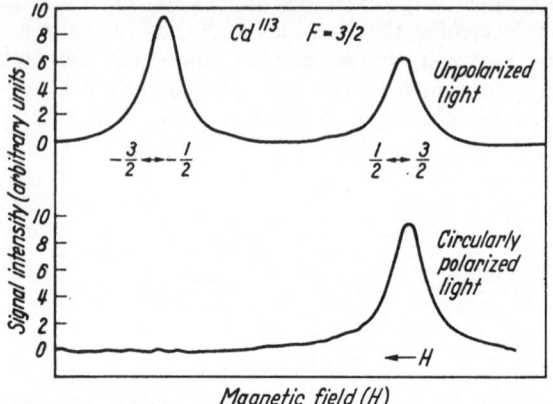

Fig. 23. Intermediate-field Zeeman resonances in the $F = 3/2$ state of Cd^{113} as observed with unpolarized and σ^+ circulary polarized exciting light. Note that the field increases from right to left

exciting light both transitions $m_F = +3/2$ to $m_F = +1/2$ and $m_F = -1/2$ to $m_F = -3/2$. Below, with σ^+-excitation, only the transition with the

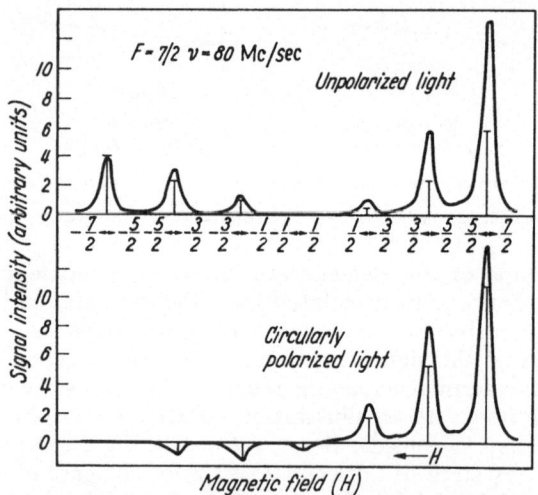

Fig. 24. Intermediate-field Zeeman resonances in the $F = 7/2$ state of Cd^{109} as observed with unpolarized and σ^+ circulary polarized exciting light. The vertical bars indicate the theoretical intensities normalized to the high-field resonance. Note that the field increases from right to left.

higher resonance frequency can be seen[*]. Then from Fig. 22 it is seen that the hfs of Cd^{113} is inverted, i. e., that the A-factor is negative. From that it can be concluded that Cd^{113} has a negative magnetic moment. Fig. 24 shows a similar investigation with Cd^{109}. It also has an inverted hfs.

[*] Since the magnetic field strength increases from right to left the resonances in Figs. 23 and 24 correspond to the case of constant magnetic field with the frequency increasing from left to right.

The difficulty can arise that two radioactive isotopes with the same spin and roughly the same hfs originate through the production of the

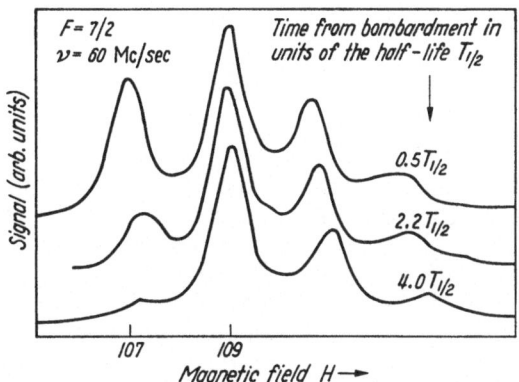

Fig. 25. Part of the $F = 7/2$ Zeeman spectrum for Cd^{107} and Cd^{109} showing the decay of the Cd^{107} resonances. The numbers 107 and 109 indicate the isotopic assignment

radioactivity. This was the case with the Cd^{107} experiment [99] ((p, n) reactions in naturally-occurring Ag^{107} and Ag^{109} led to Cd^{107} and Cd^{109}). A spin determination in weak field is then quite uncertain. It was carried out in this case in interme-diate-field where the reso-nances of the individual iso-topes could be resolved. It was possible to assign each resonance to the correct iso-tope because of the differ-ence in the half-lives (Cd^{107} (6.7 h), Cd^{109} (470 d)). Fig. 25 shows part of the Zee-man spectrum in interme-diate-field for the $F = 7/2$ level. One sees how the com-ponent ascribed to Cd^{107} markedly decreases with time relative to the Cd^{109} sig-

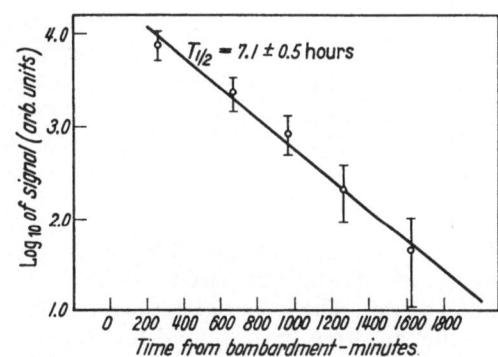

Fig. 26. Decay curve for one of the Cd^{107} resonances. The Cd^{107} signal has been normalized to one of the Cd^{109} resonances in order to reduce the effect of instrumental fluctuations

nal. In this case it was not only possible to assign the signals to two different isotopes but also to check the half-life of Cd^{107} and thus make a positive identification of the isotope. Fig. 26 shows the decrease in the Cd^{107} resonance signal (normalized to the Cd^{109} signal) as a function of time. The half-life was found to be $T_{1/2} = 7.1(5)$ h in comparison with the value $T_{1/2} = 6.7$ h [104] measured with nuclear physics techniques. An investigation similar to the one just described was carried out for the Cd isotopes Cd^{113m}, Cd^{115} and Cd^{115m} as well as for Zn^{65} [100−103]. For Zn it was more feasible to observe hfs transitions in zero-field since the two isotopes in the resonance cell, Zn^{65} and Zn^{67}, have the same nuclear spin $I = 5/2$ and a similar hfs splitting. It is expected that the investigation of radioactive nuclei of the group 2b will yield even more results — results which will facilitate the study of nuclear properties under the influence of the addition of neutron pairs.

VI. Level crossing experiments

In a double resonance experiment the change in the angular distribution and polarization of the resonance radiation occurs because an rf-magnetic field causes the mixing of the excited levels. It is, however, thinkable that two excited levels of an atom which are resolved because of the hfs interaction can be made to move closer to one another and eventually to cross by the application of a magnetic field. This phenomenon is known as "level crossing". The experiments of Hanle [105] on zero-field crossings demonstrated that the degeneracy of the two states results in a change in the polarization of the resonance light. Level crossings in magnetic field $H_0 \neq 0$ were first seen by Colegrove, Franken, Lewis and Sands in the He I-spectrum [106]. The technique has been applied to many other atoms since then. Franken [107] as well as

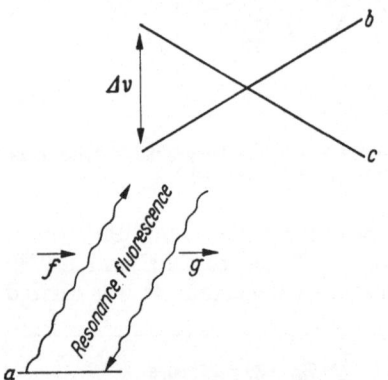

Fig. 27. Energy level diagram. The excited states b and c are separated in zero magnetic field by the amount $\Delta\nu$ due to fine or hyperfine structure interactions and cross at some specific value of the field. In the resonance fluorescence process, photons of polarization $\vec{f}$ are absorbed and those of polarization $\vec{g}$ are re-emitted

Rose and Carovillano [108] have given a theoretical explanation of the effect which is based on the work of Breit [109] on resonance fluorescence. The essential expressions will be given here[*], [**].

For an atom with ground levels m, m' and excited levels μ, μ' the rate R at which photons with polarization f are absorbed and photons with polarization g are emitted is given by the Breit formula

$$R(\boldsymbol{f}, \boldsymbol{g}) = k \sum_{\mu\mu'mm'} \frac{f_{\mu m} f_{m\mu'} g_{\mu' m'} g_{m' \mu}}{1 - 2\pi i \tau \nu(\mu, \mu')} \tag{32}$$

where k is a constant which depends on the photon density and the geometry of the excitation, $f_{\mu m}$, $g_{\mu' m'}$, are matrix elements of the form $f_{\mu m} = \langle \mu | \boldsymbol{f} \cdot \boldsymbol{r} | m \rangle$, τ is the lifetime of the excited levels and $\nu(\mu, \mu')$ is the energy difference (in frequency units) between the excited levels. If one considers the special case of one ground level a and two excited levels b and c (see Fig. 27) which are completely resolved ($2\pi \tau \nu(b, c) \gg 1$),

---

[*] Series [110] has described anticrossings in optical resonance fluorescence.
[**] The treatment here follows that of Franken [107].

then Eq. (32) becomes

$$R(f, g) = R_0 = k\{|f_{ab}|^2 |g_{ba}|^2 + |f_{ac}|^2 |g_{ca}|^2\}. \tag{33}$$

This is the well-known relation for resonance fluorescence without interference terms. If the levels b and c lie near one another ($2\pi \tau \nu (b,c) \lesssim 1$) Eq. (32) becomes

$$R(f, g) = R_0 + \frac{A}{1 - 2\pi i \tau \nu (b, c)} + \frac{A^*}{1 + 2\pi i \tau \nu (b, c)} = R_0 + S \tag{34}$$

with $A = f_{ba} f_{ac} g_{ca} g_{ab}$. The term S may be rewritten

$$S = \frac{A + A^*}{1 + 4\pi^2 \tau^2 \nu^2 (bc)} + \frac{(A - A^*) 2\pi i \tau \nu (bc)}{1 + 4\pi^2 \tau^2 \nu^2 (bc)}. \tag{35}$$

If the matrix product A is real the second term disappears and

$$S = \frac{2A}{1 + 4\pi^2 \tau^2 \nu^2 (bc)}. \tag{36}$$

That is a Lorentz curve with full width at half maximum $\Delta \nu (b, c) = 1/\pi \tau$, twice the natural width of the levels b, c. For a pure imaginary matrix product A S becomes

$$S = \frac{4A \pi i \tau \nu (b, c)}{1 + 4\pi^2 \tau^2 \nu^2 (b, c)}. \tag{37}$$

This is a dispersion curve. For complex A the signal has a form which is intermediate between these two extremes. This property — real, imaginary, or complex — of A depends on the orientation and polarization of the exciting and emitted light.

1. Level crossing in the $^3P_1$-state of Cd and Hg isotopes with $I = 1/2$

As an example of a level crossing experiment one may consider the measurement of the hfs splitting of the two odd stable Cd isotopes Cd¹¹¹ and Cd¹¹³ (both $I = 1/2$) [111]. Fig. 28 depicts the energy level diagram for Cd¹¹³. As is seen the levels $(F = 1/2, m_F = -1/2)$ and $(F = 3/2, m_F = +1/2)$ cross each other (here the weak field notation is

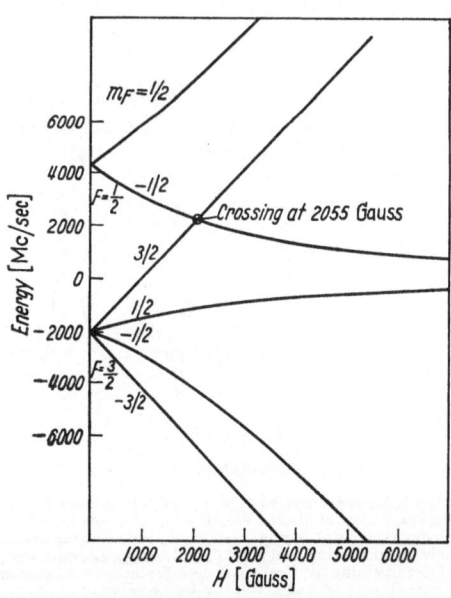

Fig. 28. Zeeman effect of the hyperfine structure of the (5s 5p) $^3P_1$ state of Cd¹¹³

used). Using the Breit-Rabi formula the magnetic field H_c at which the levels cross can be calculated to be

$$H_c = \frac{A\,(^3P_1)}{(g_J + 1/2\,g'_I)\,\mu_0}\,. \tag{38}$$

This equation allows the calculation of the A-factor from the crossing field if g_J and g'_I are known.

The experimental set up (see Fig. 29) is very similar to a double resonance experiment. Because of the magnetic field at the resonance cell the photomultiplier is set up at some distance from the cell and is provided with a light pipe. H_c is measured with proton resonance. The constancy of the lamp intensity over the crossing region must be assured in order to avoid line shifts.

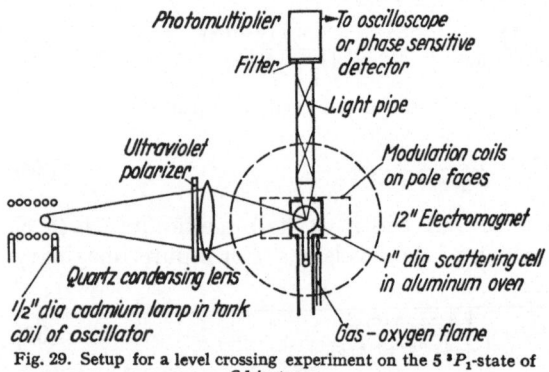

Fig. 29. Setup for a level crossing experiment on the 5 $^3P_1$-state of Cd isotopes

As in all level crossing experiments the geometry is determined by the signal form which one wishes to obtain. The magnetic field H_0 is perpendicular to the plan of the paper (Fig. 29). The exciting light is linearly polarized perpendicular to the magnetic field. The result for the matrix product A for the $^3P_1$-state with $I = 1/2$ is

$$A = (1/12)\ [\cos(2\Theta) - i\sin(2\Theta)] \tag{39}$$

where Θ is the angle between the direction of polarization of the exciting and the resonance light. The signal is then a Lorentz curve for $\Theta = 90°$ and a dispersion curve for $\Theta = 45°$ (135°). Both of these line shapes and

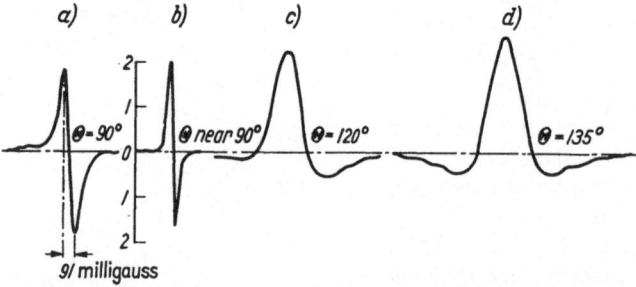

Fig. 30. Recorder traces of the crossing point effect for Cd¹¹², using phase-sensitive detection at 30 cps, with time constants of about 1 sec. The signal shown is the derivative of the true line shape, providing the field modulation is sufficiently small. In (a) the incoming and scattered light cones have been adjusted to give an effective scattering angle of 90°, and the upper and lower peaks are of equal size, while in (b) the lower half of the incoming light cone has been blocked out to give an effective scattering angle greater than 90°, with a consequent slight admixture of the dispersion-type line shape, and an asymmetry of the observed derivative. In (c) the scattering angle is 120° and in (d) 135°, where the signal is the derivative of a pure dispersion-type curve. The various apparent linewidths shown are due to different field-sweep rates

some intermediate forms were observed experimentally. They are shown in Fig. 30. Note that the phase-sensitive detection used to record the signals displays the derivative of the line shape.

The line width when expressed in units of magnetic field is dependent on the angle at which the levels cross as well as on the lifetime of the levels. For the case described the line width at $\Theta = 90°$ amounted to

$$\Delta H = (dH/d\nu(b, c))_{H_e} \cdot (1/\pi \tau) = 45 \cdot 10^{-3} \text{ Gauss} \qquad (40)$$

at a crossing field $H_e = 2055$ Gauss. The resolution of the signal was then about $2 \cdot 10^{-5}$. The following other isotopes with $I = 1/2$ were also investigated in this manner: Hg199 [112, 113, 91], Hg197 [114, 91] and Hg195 [115].

2. Level crossing experiments for Zn, Cd and Hg isotopes with $I > 1/2$

The level crossing experiments on the $^3P_1$ states of $I = 1/2$ isotopes which were described above were extended to isotopes with $I > 1/2$ (the first such experiment was the measurement of Hg197m $(I = 13/2)$ [91]). From the crossings one obtains A- and B-values which are just as exact as those obtained in a double resonance experiment at zero-field. Later Zn65 and Zn67 [117] as well as Cd107 and Cd109 [116] and Hg203 [129] were measured with the level crossing method. Since these five nuclei have $I = 5/2$ the case of Cd107[116] will serve to illustrate all of them. Fig. 31 shows the hfs energy level diagram of the 5 $^3P_1$-state of Cd107. Most of the crossings are "foldover" crossings: crossings of sublevels which originate in the same F-level. Fig. 32 is a magnification of the interesting part of Fig. 31. As is seen there are $(2I-1)$ crossings for $\Delta m_F = 2$ and $(2I-1)$ crossings for $\Delta m_F = 1$ (this is always the case for $J = 1$ and $I > 1/2$).

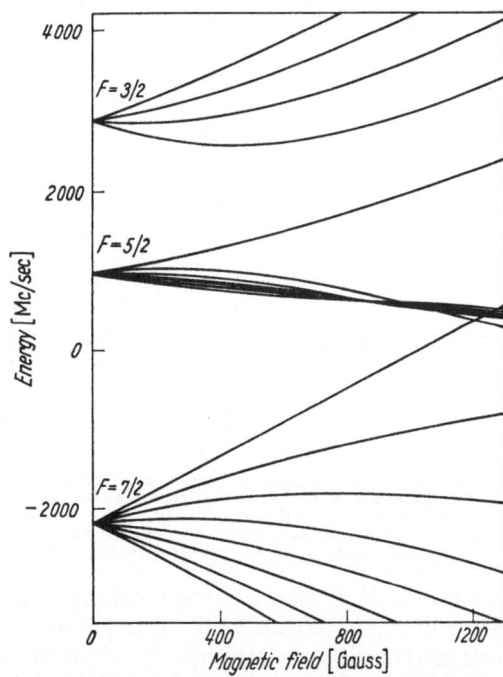

Fig. 31. Zeeman effect of the (5s 5p) $^3P_1$ state of Cd107 $(I = 5/2)$

Only these crossings give signals. The single $\Delta F = 1$ crossing is clearly the sharpest. And its signal also happens to be the strongest. The $\Delta F = 1$ crossing provides an exact value of A/g_J since, as can be shown, its position does not depend on B. Of the other crossings the $\Delta m_F = 2$ ones are the more interesting because they are easier to observe. If the incoming light beam, the outgoing resonance light beam and the static magnetic field are oriented perpendicular to each other the $\Delta m_F = 2$ signal has the Lorentz form even without the use of any polarization filter. The "foldover" crossings are very much broader than the $\Delta F = 1$ crossing because of the smaller relative inclination of the crossing levels. The "foldover" signals are also smaller. They have, however, a strong dependence on the B-factor and are therefore suited for an exact determination of B. The experimental setup corresponds to that which was

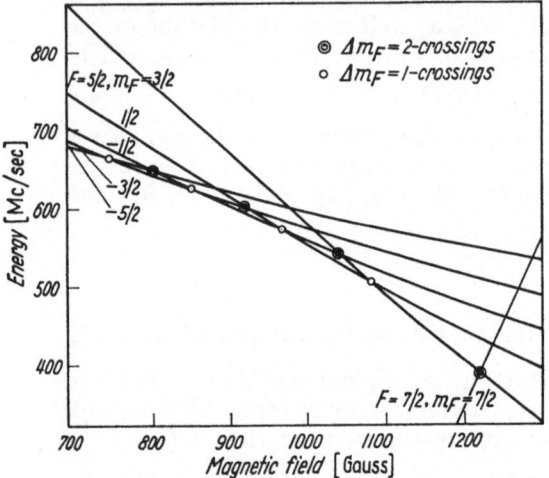

Fig. 32. Details of the level crossings in the $(5s\,5p)\,^3P_1$ state of Cd^{107}

illustrated in part 1 of this section. Fig. 33 shows the signals observed: at the left the $\Delta F = 1$, $\Delta m_F = 2$ crossing, at the right the "foldover" $\Delta F = 0$, $\Delta m_F = 2$ crossings.

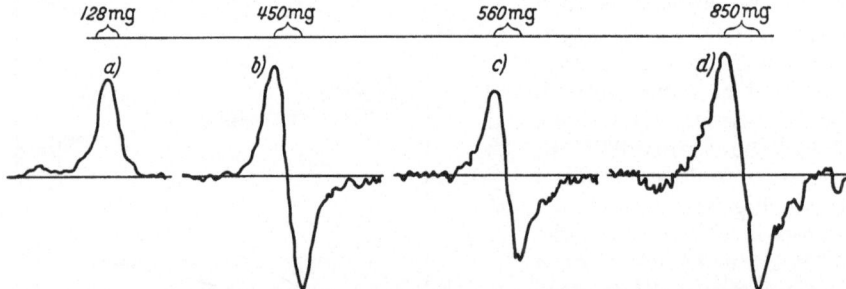

Fig. 33. Observed level crossings for Cd^{107}. (a) is an oscilloscope trace of the main (7/2, 7/2), (5/2, 3/2) crossing. (b), (c), and (d) are recorder traces with phase-sensitive detection of the $\Delta F = 0$ crossings: (5/2, 3/2), (5/2, —1/2); (5/2, 1/2), (5/2, —3/2); and (5/2, —1/2), (5/2, —5/2), with time constants of 1, 3, and 10 sec, respectively. The vertical scale is arbitrary in each case. (a) has a Lorentzian line shape, while (b), (c), and (d) have line shapes, due to phase-sensitive detection, which are the derivative of a Lorentzian

The state of affairs is more difficult for Hg$^{197\,m}$ than for Cd since the hfs splitting is considerably larger. With the excitation technique which was used in the scanning experiments it was possible to obtain a maximum intensity of the exciting light in the region of the crossing. The large magnetic field H_e required for the observation of the crossings led to a somewhat broadened line due to inhomogeneities in the magnetic field. Instead of lock-in detection a bridge circuit which compared the intensities of the exciting and resonance light was employed.

The special advantage of the level crossing method was clearly evident in this Hg[197m] experiment. The production of the rf-field strengths required to induce transitions in short-lived states is no easy matter at these high frequencies. In addition the level crossing signals are large compared with the rf signals in double resonance experiments. This is so because all atoms in the levels concerned contribute to the signal (as opposed to just those which are influenced by the rf-field). For these reasons level crossing experiments play a significant roll in measuring the structure of short-lived atomic states.

3. Summary of results for the hfs splitting of the $^3P_1$-states of Zn, Cd and Hg isotopes

In Table 5 are compiled the results for the hfs splitting of the $^3P_1$-states of the stable and radioactive Zn, Cd and Hg isotopes which have been investigated up to now. The letters signify: a, double resonance in zero- or weak-field; b, double resonance in intermediate- or strong-field; c, level crossing. As in the case of Tables 2 and 3, values obtained with other methods such as atomic beam resonance or optical spectroscopy are not included in Table 5. Such measurements were mentioned to some degree in the text. The magnetic moments of the radioactive nuclei were calculated by the authors of the original papers from the known ratio g'_I/A for a stable isotope. In this process the uncertainty was in general enlarged to allow for the possibility of a hfs anomaly. The Q-values are given without the Sternheimer correction. Those values which have been corrected for second order effects are indicated with *. The uncorrected values are also given in some cases to show the size of the correction.

VII. Summary

The method of optical double resonances offers, through the combination of optical and rf methods, the possibility of investigating with high precision the structure of excited atomic states. In this article was discussed that portion of the experiments which served to determine nuclear data from the hfs of excited states. The investigation of a complete series of terms of an atom yields additional information about atomic properties. This knowledge is extended still more through the measurement of lifetimes and g_J-factors with double resonance.

A large number of methods for the investigation of excited atomic states has not been mentioned since they have not yet been used for the determination of nuclear moments. With electron bombardment one can reach states which cannot be populated optically. The spectra of ions have also been investigated [118]. Excitation with modulated light [119, 120] or pulsed electron bombardment [121] leads to additional

Table 5. *Compilation of hfs measurements of isotopes of group 2 b elements*

Nucleus	State	Meth.	$A_{Mc/s}$	$B_{Mc/s}$	I^{**}	μ_I/μ_K^{**}	$Q_{10^{-24}\,cm^2}^{***}$	Ref.
Zn65	$4\,^3P_1$	a	+ 535,117(2)	+ 2,445(4)	5/2			[102]
		a	+ 535,163(2)*	+ 2,870(5)*		+0,7688(6)	−0,024(2)	[102]
		c		+ 2,867(12)*				[117]
Zn67	$4\,^3P_1$	a	+ 608,99(5)	− 19,37(9)				[97]
		a	+ 609,208(2)	− 19,331(7)			+0,18(2)	[102]
		a	+ 609,086(2)*	− 18,782(8)*				[102]
		c		− 18,770(12)*				[117]
Cd107	$5\,^3P_1$	b	− 854,2(1,0)	−166(3)	5/2		+0,77(10)	[99]
		c	− 853,543(6)*	−163,279(5)*		−0,6162(8)		[116]
Cd109	$5\,^3P_1$	b	− 1148,6(2,0)	−167,3(2,0)	5/2		+0,78(10)	[86]
		c	− 1148,784(7)*	−165,143(5)*		−0,8286(15)		[116]
Cd111	$5\,^3P_1$	a	− 4123,81(1)					[96]
	$5\,^1P_1$	c	− 186(4)					[132]
Cd113	$5\,^3P_1$	a	− 4313,86(1)					[96]
Cd113m	$5\,^3P_1$	b	− 686,0425(8)*	+169,047(9)*	11/2	−1,08885(13)	−0,79(10)	[100, 101]
Cd115	$5\,^3P_1$	b	− 4484(2)		1/2	−0,6469(3)		[103]
Cd115m	$5\,^3P_1$	b	− 657,6(6)	+131(6)		−1,0437(10)	−0,61(8)	[103]
Hg193	$6\,^3P_1$	a	+16600(1100)		1/2	+0,562(35)		[134, 129]
Hg193m	$6\,^3P_1$	c	− 2399,69(6)	−724,8(90,0)	13/2		+1,3(3)+	[115]
Hg195	$6\,^3P_1$	c	+15813,46(23)			+0,538860(16)		[115]
Hg195m	$6\,^3P_1$	c	− 2368,04(8)	−782,45(86)		−1,04903(13)	+1,40(12)+	[115]
Hg197	$6\,^3P_1$	b	+15394(30)†					[90]
		c	+15388,9(4,5)					[114]

Hg^{197m}	$6\,^3P_1$	b,c	+15387,1(5,3)			[91]
		a	+15390,91(1)			[94]
		a	+15392,66(15)*	−901(13)	+1,61(13)+	[94]
Hg^{199}	$6\,^3P_1$	b,c	− 2328,8(1,7)			[91]
		a	− 2328,89(84)	−902,9(5,4)		[133]
		b	+14900(390)†			[95]
		b	+14733,3(15,0)			[73]
Hg^{201}	$6\,^3P_1$	b,c	+14750,7(5,0)			[91]
		a	+14752,37(1)			[94]
		a	+14754,04(14)*	−283(19)		[94]
		b	− 5437,1(15,0)			[73]
		a	− 5454,569(3)	−280,107(5)	+0,50(4)+	[14]
Hg^{203}	$6\,^3P_1$	c	+ 4991,37(1)	−255(1)	+0,46(4)+	[129]

a Zero- or weak-field measurement.
b Intermediate- or strong-field measurement.
c Level crossing.

* Corrected for second order effects.
** Spins or moments determined by double resonance.
*** Without Sternheimer correction.
† Using 0,001 cm⁻¹ = 29,97930(3) MHz [130],
+ Q is obtained from Q^{201} [83] and the ratios of B to B^{201}.

experiments. Furthermore one can use a metastable state as the initial
state for optical excitation [122].

Also, the double resonance method is not confined to elements of
the first two main groups of the periodic system. Successful studies of the
excited states of xenon [123], thallium [124], chromium [125], lead [126],
iron [127] and copper [128] have been reported.

Thus it is to be assumed that also in the future a great number
of atomic and nuclear properties will be determined with optical-rf-
spectroscopic measurements on excited atomic states.

The author is grateful to his colleagues in the "I. Physikalisches
Institut der Universität Heidelberg" for many worthwhile discussions
during the writing of this article. Particular thanks go out to Dr. Robert
A. Haberstroh, presently in Heidelberg as a National Science Founda-
tion Postdoctoral Fellow, for many comments and suggestions as well as
for the translation of the manuscript.

References

[1] Haxel, O., J. H. D. Jensen, and H. E. Suess: Phys. Rev. 75, 1766 (1949)
    Goeppert-Mayer, M.: Phys. Rev. 75, 1969 (1949)
    Goeppert-Mayer, M., and J. H. D. Jensen: Elementary Theory of Nuclear
    Shell Structure. New York: Wiley 1955
[2] Bohr, A.: Kgl. Danske Videnskab. Selskab., Mat. Fys. Medd. 26, No. 1
    (1952)
    Bohr, A., and B. R. Mottelson: Kgl. Danske Videnskab. Selskab., Mat.
    Fys. Medd. 27, No. 16 (1953)
    Nilsson, S. G.: Kgl. Danske Videnskab. Selskab., Mat. Fys. Medd. 29, No. 16
    (1955)
[3] de-Shalit, A., and I. Talmi: Nuclear Shell Theory. New York: Academic
    Press 1963
[4] Kisslinger, L. S., and R. A. Sorensen: Revs. Modern Phys. 35, 853 (1963)
    Arima, A., and H. Horie: Progr. Theoret. Phys. (Kyoto) 12, 623 (1954)
    Noya, A., A. Arima, and H. Horie: Progr. Theoret. Phys. (Kyoto) Suppl. 8,
    33 (1958)
[5] Lindgren, I.: Table of Nuclear Spins and Moments, "Perturbed Angular
    Correlations", p. 379. Amsterdam: North-Holland 1964
    Mack, J. E.: Revs. Modern Phys. 22, 64 (1950)
    Laukien, G.: Handbuch der Physik, Vol. 38/1, p. 338
    Townes, C. H.: Handbuch der Physik, Vol. 38/1, p. 433
[6] Kopfermann, H.: Nuclear Moments. New York: Academic Press 1958
[7] Rabi, I., J. Zacharias, S. Millman, and P. Kusch: Phys. Rev. 53, 318
    (1938)
[8] Ramsey, N. F.: Molecular Beams. Oxford: Clarendon Press 1956
[9] Nafe, J. E., and E. B. Nelson: Phys. Rev. 73, 718 (1948)
[10] Lurio, A., G. Weinreich, C. W. Drake, V. W. Hughes, and J. A. White:
     Phys. Rev. 120, 153 (1960)
[11] Rabi, I.: Phys. Rev. 87, 379 (1952)
[12] Kastler, A., et J. Brossel: Compt. Rend. 229, 1213 (1949)
[13] Brossel, J., and F. Bitter: Phys. Rev. 86, 308 (1952)
[14] Kohler, R. H.: Phys. Rev. 121, 1104 (1961)
[15] Bitter, F.: Appl. Optics 1, 1 (1962)
[16] Dehmelt, H. G.: Phys. Rev. 103, 1125 (1956)
[17] Pebay-Peyroula, J. C.: J. Phys. Rad. 20, 721 (1959)

[18] LAMB, W. E.: Phys. Rev. **105**, 559 (1957)
[19] DEHMELT, H. G.: Phys. Rev. **109**, 381 (1958)
[20] KASTLER, A., et J. BROSSEL: Cahiers de Physique **65** (1956)
[21] — J. Opt. Soc. Am. **47**, 460 (1957)
[22] *Colloques Internationaux LXXXV*, Sur la Résonance Magnétic, Paris 1958
[23] SERIES, G. W.: Radio-Frequency Spectroscopy of Excited Atoms. Repts. Progr. in Phys. **22**, 280 (1959)
[24] KOPFERMANN, H.: Über optisches Pumpen an Gasen. Sitzber. Heidelb. Akad. Wiss., Math.-naturw. Kl. **1960**, 3 Abh.
[25] KASTLER, A., J. BROSSEL, T. SKALINSKY, C. COHEN-TANNOUDJI, A. D. MAY, and J. WINTER: Topics in Radio-Frequency Spectroscopy, Course 17 of the Proceedings of the International School of Physics „Enrico Fermi". New York: Academic Press 1962
[26] *International Conference on Optical Pumping*, Heidelberg, April 26—28, 1962.
[27] BARNES, R., and W. SMITH: Phys. Rev. **93**, 95 (1954)
[28] SENITZKY, B., and I. I. RABI: Phys. Rev. **103**, 315 (1956)
[29] PHILLIPS, M.: Phys. Rev. **103**, 322 (1956)
[30] JUDD, B. R.: Lectures at the Seventh Brookhaven Conference on Molecular Beams and Atomic Resonance, Uppsala 1964
BAUCHE, J., and B. R. JUDD: Proc. Phys. Soc. **83**, 145 (1964)
[31] SCHÜSSLER, H. A.: Z. Physik. **182**, 289 (1965)
[32] STERNHEIMER, R.: Phys. Rev. **80**, 102 (1950)
[33] — Phys. Rev. **84**, 244 (1951)
[34] — Phys. Rev. **86**, 316 (1953)
[35] — Phys. Rev. **95**, 736 (1954)
[36] — Phys. Rev. **105**, 158 (1957)
[37] — Phys. Rev. **127**, 1220 (1962)
[38] WATSON, R. E., and A. J. FREEMAN: Phys. Rev. **131**, 250 (1963)
[39] STERNHEIMER, R.: private communication 1963—1964
[40] WAGNER, R.: Heidelberg (1960) unpublished
[41] KRÜGER, H.: Z. Physik **141**, 43 (1955)
[42] ALTHOFF, K. H.: Z. Physik **141**, 33 (1955)
[43] BROSSEL, J., B. CAGNAC, et A. KASTLER: J. Phys. Rad. **15**, 6 (1954)
[44] BESSET, C., J. HOROWITZ, A. MESSIAH, and J. WINTER: J. Phys. Rad. **15**, 251 (1954)
[45] BUCKA, H., H. KOPFERMANN, and E. W. OTTEN: Ann. Phys. 7 F, **4**, 39 (1959)
[46] — Z. Physik **151**, 328 (1958)
[47] BLOCH, F., and A. SIEGERT: Phys. Rev. **57**, 522 (1940)
[48] RITTER, G. J., and G. W. SERIES: Proc. Roy. Soc. A. **238**, 473 (1957)
[49] DODD, J. N., and R. W. N. KINNEAR: Proc. Phys. Soc. **75**, 51 (1960)
[50] FOX, W. N., and G. W. SERIES: Proc. Phys. Soc. **77**, 1141 (1961)
[51] BUCKA, H., H. KOPFERMANN, M. RASIWALA, and H. SCHÜSSLER: Z. Physik **176**, 45 (1963)
[52] — —, and A. MINOR: Z. Physik **161**, 123 (1961)
[53] — G. ZU PUTLITZ, and R. RABOLD: Submitted to Z. Physik
[54] — — 7th Brookhaven Conference on Molecular Beams and Atomic Resonance, Uppsala, Sweden, 1964
[55] ZU PUTLITZ, G., and A. SCHENCK: Z. Phys. **183**, 428 (1965)
[56] SAGALYN, P. L.: Phys. Rev. **94**, 885 (1954)
[57] KRÜGER, H., and K. SCHEFFLER: In [22]
[58] RITTER, G. J., and G. W. SERIES: Proc. Phys. Soc. A, **68**, 450 (1955)
[59] BUCKA, H., H. KOPFERMANN, and J. NEY: Z. Physik **159**, 49 (1960); **167**, 375 (1962)
[60] MEYER-BERKHOUT, U.: Z. Physik **141**, 185 (1955)
[61] KRÜGER, H., and U. MEYER-BERKHOUT: Naturwiss. **42**, 94 (1955)
[62] ALTHOFF, K., u. H. KRÜGER: Naturwiss. **41**, 368 (1954)
[63] BUCKA, H., H. KOPFERMANN u. E. W. OTTEN: Naturwiss. **45**, 620 (1958)
[64] —, and G. V. OPPEN: Ann. Phys. 7 F, **10**, 119 (1962)
[65] —, and H. H. NAGEL: Ann. Phys. 7 F, **8**, 329 (1961)

[66] Bucka H., H. Kopfermann u. G. zu Putlitz: Z. Physik 165, 72 (1961)
[67] — — u. G. zu Putlitz: Z. Physik 168, 542 (1962)
[68] zu Putlitz, G.: Ann. Phys. 7 F, 11, 248 (1963)
[69] — Z. Physik 175, 543 (1963)
[70] Casimir, H. B. G.: On the interaction between Atomic Nuclei and Electrons. Teylors, Tweede Genotshap 11, 1936
[71] Wolfe, H. C.: Phys. Rev. 41, 443 (1936)
[72] Lurio, A., M. Mandel, and R. Novick: Phys. Rev. 126, 1758 (1962)
[73] Sagalyn, P. L., A. C. Melissinos, and F. Bitter: Phys. Rev. 109, 375 (1958)
[74] Walchi, H. E., and T. J. Rowland: Phys. Rev. 102, 1334 (1956)
[75] Perl, M. L., I. I. Rabi, and B. Senitzki: Phys. Rev. 98, 611 (1955)
[76] Buck, P., and I. I. Rabi: Phys. Rev. 107, 1291 (1957)
[77] Senitzki, B., and I. I. Rabi: Phys. Rev. 103, 315 (1956)
[78] Lurio, A., R. de Zafra, and R. Goshen: Phys. Rev. 134, A 1198 (1964)
[79] Byron, F. W., M. N. McDermott, R. Novick, B. W. Perry, and E. B. Saloman: Phys. Rev. 134, A 47 (1964)
[80] Schwartz, C.: Phys. Rev. 105, 173 (1957)
[81] Hughes, D. S., and C. Eckart: Phys. Rev. 36, 694 (1930)
[82] Lurio, A.: Phys. Rev. 126, 1768 (1962)
[83] McDermott, M. N., and W. Lichten: Phys. Rev. 119, 134 (1960)
[84] — — Phys. Rev. 120, 469 (1960)
[85] Jackson, D. A., and D. Hong Tuan: Proc. Phys. Soc. 280, A 323 (1964)
[86] McDermott, M. N., and R. Novick: Phys. Rev. 131, 707 (1963)
[87] Stroke, H. H., R. J. Blin-Stoyle, and V. Jaccarino: Phys. Rev. 123, 1326 (1961)
[88] Blaise, J., and H. Chantrel: J. Phys. Rad. 18, 193 (1957)
[89] Melissinos, A. C., and S. P. Davis: Phys. Rev. 115, 130 (1959)
[90] Melissinos, A. C.: Phys. Rev. 115, 126 (1959)
[91] Hirsch, H. R.: J. Opt. Soc. Am. 51, 1192 (1961)
[92] Walter, W. T., and H. H. Stroke: Bull. Am. Phys. Soc. 9, 452 (1964)
[93] Bitter, F.: Appl. Opt. 1, 1 (1962)
[94] Stager, C. F.: Phys. Rev. 132, 275 (1963)
[95] Bogle, G. S., J. N. Dodd, and W. L. McLean: Proc. Phys. Soc. B 70, 796 (1957)
[96] Lacey, R. F.: Ph. D. Thesis MIT 1959 (unpublished). Values can be found in [111]
[97] Böckmann, K., H. Krüger, and E. Recknagel: Ann. Phys. 6 F, 20, 250 (1957)
[98] Nierenberg, W.: Ann. Rev. Nuc. Sci. 7, 349 (1957)
[99] Byron, F. W., M. N. McDermott, and R. Novick: Phys. Rev. 132, 1181 (1963)
[100] McDermott, M. N., R. Novick, and B. W. Perry: Bull. Am. Phys. Soc. 8, 262 (1963); 8, 345 (1963)
[101] Byron, F. W., M. N. McDermott, R. Novick, B. W. Perry, and E. B. Saloman: Phys. Rev. 136, B 1654 (1964)
[102] — — — — — Phys. Rev. 134, A 47 (1964)
[103] McDermott, M. W., R. Novick, B. W. Perry, and E. Saloman: Phys. Rev 134, B 25 (1964)
[104] Helmholz, A. C.: Phys. Rev. 60, 415 (1941)
[105] Hanle, W.: Z. Physik 30, 93 (1924); Ergeb. exakt. Naturwiss. 4, 214 (1925)
[106] Colegrove, F. D., P. A. Franken, R. R. Lewis, and R. H. Sands: Phys. Rev. Letters 3, 420 (1959)
[107] Franken, P. A.: Phys. Rev. 121, 508 (1961)
[108] Rose, M. E., and R. L. Carovillano: Phys. Rev. 122, 1185 (1961)
[109] Breit, G.: Revs. Modern Phys. 5, 91 (1933)
[110] Series, G. W.: Phys. Rev. Letters 11, 13 (1963)
[111] Thaddeus, P., and R. Novick: Phys. Rev. 126, 1774 (1962)
[112] Dodd, J. N.: Proc. Phys. Soc. 77, 669 (1961); 78, 65 (1961)
[113] Hirsch, H. R.: Bull. Am. Phys. Soc. 5, 274 (1960)

[114] HIRSCH, H. R., and C. V. STAGER: J. Opt. Soc. Am. **50**, 1052 (1960)
[115] SMITH, W. W.: Phys. Rev. **137**, A 330 (1965), Bull. Am. Phys. Soc. **8**, 9 (1963)
[116] THADDEUS, P., and M. N. McDERMOTT: Phys. Rev. **132**, 1186 (1963)
[117] LANDMAN, A., and R. NOVICK: Phys. Rev. **134**, A 56 (1964)
[118] BARRAT, M.: Compt. Rend. **259**, 1504 (1964)
[119] CORNEY, A., and G. W. SERIES: Proc. Phys. Soc. **83**, 207 (1964)
[120] ALEXANDROV, E. B.: Optics and Spectroscopy (SSSR) **14**, 436 (1963)
[121] DEHMELT, H. G.: Phys. Rev. **103**, 1125 (1956)
       PEBAY-PEYROULA, J. C., J. BROSSEL, and A. KASTLER: Compt. Rend. **244**, 57 (1957); **245**, 840 (1957)
       PEBAY-PEYROULA, J. C., and O. NEDELEC: Quantum Electronics III, Edited by P. GRIVET and N. BLOEMBERGEN, Paris, Dunod Ed. and New York: Columbia Univ. Press, 1964, p. 287 where a summary of measurements obtained by electron impact can be found
       NEDELEC, O., M. N. DESCHIZEAUX, et J. C. PEBAY-PEYROULA: Compt. Rend. **257**, 3130 (1963)
[122] BROSSEL, J., et C. JULIENNE: Compt. Rend. **242**, 2117 (1956)
[123] KENT-ANDERSON, P.: Phys. Rev.; to be published
[124] GALLAGHER, A., and A. LURIO: Phys. Rev. **136**, A 87 (1964)
[125] BUDICK, B., R. I. GOSHEN, and S. MARCUS: Bull. Am. Phys. Soc. **9**, 448 (1964)
[126] NOVICK, R., B. W. PERRY, and E. B. SALOMAN: Bull. Am. Phys. Soc. **9**, 625 (1964)
[127] OTTEN, E. W., and R. WAGNER: Private communication
[128] ZU PUTLITZ, G., and J. KOWALSKI: Private communication
[129] REDI, O.: Quarterly Progr. Rep. No. 74, Research Lab. of Electronics, MIT (July 1964), p. 43
       TOMLINSON III, W. J., and H. H. STROKE: Nucl. Phys. **60**, 614 (1964)
[130] COHEN, E. R., J. W. M. DU MOND, T. W. LAYTON, and J. S. ROLLET: Revs. Mod. Phys. **27**, 363 (1955)
[131] LURIO, A.: Phys. Rev. **136**, A 376 (1964)
[132] LURIO, A., and R. NOVICK: Phys. Rev. **134**, A 608 (1964)
[133] BROT, C.: J. Phys. Rad. **22**, 412 (1961)
[134] WALTER, W. T., and H. H. STROKE: Bull. Am. Phys. Soc. **9**, 452 (1964)

Dr. G. ZU PUTLITZ
I. Physikalisches Institut der Universität
69 Heidelberg, Philosophenweg 12

Rigorous Symmetries of Elementary Particles[*]

H. Ekstein

Argonne National Laboratory,
Argonne, Illinois,
and
Faculté des Sciences,
Université d'Aix-Marseille

Received November 1964

Contents

1. Introduction

Considerable uncertainty concerning the dynamical laws of particle interactions exists and is likely to last. In the meantime, symmetry principles provide a powerful but incomplete set of predictive statements.

[*] Work performed under the auspices of the U. S. Atomic Energy Commission; lectures delivered at the Cargèse Summer School, 1964.

It is imperative that we exploit the reliable symmetry principles to the largest possible extent. In particular, we shall consider the space-time symmetry, i. e., the group G of all length-preserving transformations of Minkowski space — also called the full, the extended, or the complete Poincaré or Lorentz or relativity group.

In addition, we shall need the general principles of quantum mechanics, but no field theory nor analyticity assumptions. We shall, however, need postulates on asymptotic mechanics, i. e., concerning the nature of physical states in the distant past and future when they decompose into noninteracting particles. Our "axioms" are roughly those given in HAAG's 1955 paper [1] and should not be confused with the axioms of the field theorists.

From such assumptions, we may hope to "derive" all rigorous symmetries found in nature. Specifically, one may expect that symmetry theory will show the possible existence of systems which have the rigorous symmetries encountered in nature and that no others would be found — "accidents don't happen." Symmetry theory "predicted" that particles with integral and half-integral spins could exist — and these and only these were found. In the same sense, one may hope that all existing mass-degeneracies will correspond to cases that occur in the enumeration of all irreducible realizations of the Poincaré group by rays in Hilbert space.

Until 1960, the symmetry theory of the Poincaré group, although highly developed, fell considerably short of this expectation [2].

Before 1956, it was generally thought that the Poincaré group was a strict symmetry group of physics, but that other symmetries, such as charge conjugation, were needed additionally to account for the existence of particle-antiparticle doublets. The WU experiment seemed at first to refute the belief in strict symmetry under space inversion, as well as in strict charge-conjugation symmetry. LANDAU [3] and SALAM [4] simplified the situation to some extent by suggesting that the operator CP rather than P should be considered as the representative of the space-inversion element I of the Poincaré group. Indeed, it seems far more satisfactory to relate all rigorous symmetries to the symmetry of the space-time continuum than to introduce an alien, nongeometric symmetry element and to remove one of the elements of the Poincaré group.

However, this suggestion still fails to explain the existence of particle-antiparticle doublets. The irreducible projective representations of G commonly found in the literature have multiplicity (i. e., number of linearly independent states for given mass and momentum) $2s + 1$; the existence of four linearly independent states for the electron-positron doublet seems explainable only by the accidental existence of two irreducible projective representations with exactly the same mass. Furthermore, one does not understand why it is not possible to see in the laboratory the correlate of the reduction of two attached representations, i. e., why there exist no linear superpositions of electron and positron which are invariant under CP, and hence correspond to irreducible representations with multiplicity $2s + 1 = 2$.

To understand this, one had to assume separately a superselection principle which explicitly forbids a linear combination of electron and positron states to be a state vector. Until 1960, then, it was necessary to make the two assumptions of an accidental occurrence of two identical irreducible projective representations and of a superselection rule.

An attempt to simplify our picture of symmetry was made by the author in 1960 [5, 6]. It was pointed out that there existed in fact irreducible projective representations of multiplicity $2(2s + 1)$ which could be attributed to the particle-antiparticle doublets. The derivations of these representations had been achieved, but not published, by Wigner many years ago [7]. The essentially new point in these papers [5, 6] was a statement concerning the operational meaning of the time-reversal transformation, which will be discussed in the following. For the moment I want to say only that the observed superselection between particles and antiparticles follows from it deductively, as does the observed form CP of the operator $U(I)$ which represents space inversion.

My approach in these two papers is nevertheless subject to serious criticism. In the traditional application of symmetry principles to quantum mechanics, superselection rules are not considered *ab initio*, i. e., the analysis is at first carried out as though the representation space were coherent. In particular, the irreducible subspaces are considered as coherent *ab initio*. It is then not consistent to discover or to superimpose an incoherence after the operators have been determined as though there were none.

Consequently, I have reviewed the entire theory of the Poincaré group in quantum mechanics from a viewpoint which *ab initio* admits an unlimited degree of incoherence, even in the irreducible subspaces. After having considered the most general possibilities, we clearly need a principle which limits the degree of incoherence. This principle — the principle of maximal coherence — to be described in the following, says that the physical Hilbert space is as coherent as it can be without violating clear physical principles.

2. Symmetry: Operational properties

We will first give an operational definition of space-time symmetry or covariance. By "*observable*" we mean a procedure, i. e., an apparatus together with instructions. Part of these instructions refer to space-time reference points and directions. For example, an observable may be a calorimeter, together with instructions specifying its state of motion (e. g., northward velocity v with respect to the laboratory), time of measurement, and location of a mark on the calorimeter.

A new observable can be generated from one standard element by a *motion* of the apparatus, e. g. a displacement of a mark on the apparatus by a given distance in a given direction. We shall consider the following motions: space-translation, rotation and "acceleration," i. e. putting the standard apparatus on a frame in uniform translation with respect to the laboratory.

Finally, we consider time-translation as a motion. This consists in changing the delay between some event characteristic for the production of the state and an instant characteristic of the measurement.

A motion of all observables induces a permutation in the space of observables.

$$A \to LA . \tag{2.1a}$$

An "ensemble" is defined operationally by a procedure. An accelerator that is in a certain position and state of motion and that is actuated at a certain time produces a member (sample) of the ensemble.

Just as for observables, a permutation of ensembles ϱ can be obtained by changing the position, orientation, and state of motion of the producing apparatus:

$$\varrho \to L\varrho . \tag{2.1b}$$

The average of many repeated measurements of an observable A on an ensemble ϱ approaches the expectation value $\mathrm{Exp}(\varrho, A)$. Two ensembles ϱ_1 and ϱ_2 can be added simply by taking alternatingly samples obtained by either procedure; and, by the definition of the expectation value, it will be the average of the two expectation values:

$$\mathrm{Exp}(\varrho_1 + \varrho_2, A) = \frac{1}{2} \left[\mathrm{Exp}(\varrho_1, A) + \mathrm{Exp}(\varrho_2, A) \right] \quad \text{(all } A\text{)} . \tag{2.2a}$$

More generally, several ensembles ϱ_i can be weighted by fractional positive numbers $a_i (\Sigma\, a_i = 1)$ and the new ensemble may be denoted by $\Sigma\, \varrho_i\, a_i$. We have

$$\mathrm{Exp}(\Sigma\, \varrho_i\, a_i, A) = \Sigma\, a_i\, \mathrm{Exp}(\varrho_i, A) \quad \text{(all } A\text{)} . \tag{2.2b}$$

Thus, ensembles form a subset in a real linear space; it is (by definition) a convex subset since, if ϱ, σ are elements, $a\varrho + b\sigma$ are elements if and only if $a + b = 1$ and a, b are positive. Extremal points of this set are elements which cannot be further decomposed, i. e., a "decomposition" formula

$$\sigma = a\varrho_1 + b\varrho_2$$

with a, $b \neq 0$ implies $\varrho_1 = \varrho_2 = \sigma$. These extremal points are physically *states* (v. NEUMANN) or *pure states* (WEYL) and will be denoted by boldface italics.

If a simultaneous motion L of states and observables, $\varrho \to L\varrho$ and $A \to LA$, leaves all expectation values invariant

$$\mathrm{Exp}(\varrho, A) = \mathrm{Exp}(L\varrho, LA) \tag{2.3}$$

we say that L is a *symmetry*.

If the subsequent performance of two motions is defined as multiplication, the symmetries form a group, since the reversed motion (the inverse) is clearly a symmetry. The multiplication table of the group can be studied empirically. For instance, one may notice that a rotation, followed by a time-translation has the same effect, for all states and observables, as a time-translation followed by a rotation. We will assume that the group of symmetries is isomorphic to the connected group $G_{+\uparrow}$ of length-preserving transformations of Minkowski space. The association between motions and space-time transformations is obvious. Leaving aside all theory, this is the purely operational content of the principle of special relativity.

It is tempting to extend the relativistic symmetry statement to the group of *all* length-preserving transformations of Minkowski space, and we shall do so; but it is important to notice that this cannot be done by straight extension of the statement given above. The two reflections I and T (space- and time-reflection, respectively) do not, literally, induce changes in the operation of measuring instruments. Neither is it possible to reverse the direction of physical time nor to turn measuring instruments inside out. Therefore, the transformations $\varrho \to \varrho'$ and $A \to A'$ induced by I and T and by all elements of the disconnected pieces of the Poincaré group must be specified on empirical grounds, i. e., we must see which permutations of the observables and states is, in fact, a symmetry, that is, which simultaneous transformation of observables and states leaves all expectation values invariant.

In the past, little attention was given to this necessity, and for good reasons. The permutations induced by I and T are restricted in that their multiplication rules with the permutations induced by $G_{+\uparrow}$ must be isomorphic to the multiplication table of the complete Poincaré group. For the irreducible representations, this restriction (together with the physical requirements on the representatives of $G_{+\uparrow}$) is sufficient to determine the permutations induced by I and T and hence by the complete group G. As we shall see, however, not all irreducible realizations of the Poincaré group by ray operators are implemented by projective representations, and it becomes indispensable to use the physical nature of some of these permutations.

From the group multiplication table, it is clear that the automorphism induced by the space-reflection elements I must be defined operationally by requiring that in all instructions the word "left" should be changed to "right". It is only the initial point of this inversion of procedures which is questionable. For example, given a right-handed copper screw in the apparatus belonging to the observable A, should the apparatus for the measurement of A carry a left-handed copper screw, or is it necessary to "invert" also the procedure for obtaining the material which constitutes the screw? For instance, the original instructions might define "copper" as the particles which emerge from a particular slit in the shielding of a mass spectrometer fed by an accelerator. The instruction for A' then might require "inversion" of the mass spectrograph or, equivalently, require that the screw be manufactured from the particles which emerge from the mirror-symmetric slit of the same mass spectrometer.

We know now that the correct, rigorous symmetry statement requires indeed an "inversion" of the mass spectrograph which selects the particle. In other words, the permutations $A \to A'$ requires the substitution of an antiparticle for a particle. In the following, this operational definition of I will be derived rather than postulated.

There remains the time-inversion element T of G. There is, clearly, no possible direct identification between the reflection in Minkowski space and physical reversal of the time direction, and one has to have recourse to experience. (To emphasize the difference between the reflection induced by T in Minkowski space and the alterations in procedures induced by it in the space of observables, the permutation is frequently called motion-reversal rather than time-reversal.) One statement about the physical property of motion reversal is known to be indispensable from conventional analysis: motion-reversal does not reverse the sign of the energy [7]. This empirical fact determines the antiunitary nature of the time-reversal operator which is the image of the time-reversal permutation.

We shall need another property of motion reversal: intuitively speaking, we infer from experiment that the nonkinematic observables of particles composing the state should be left unchanged in transforming a state f into its time-reversed image. That is, if a state is specified by such observables as linear momenta, energies, and/or angular momenta, etc. of the individual particles, there remain other observables such as charge and baryon number which commute with the kinematic observables. These are unchanged in the time-reversed state. The justification for this statement is, of course, empirical; in the time-reversal experiments on a neutron, the comparison state consisted of a neutron, not an antineutron.

A more precise definition of the kinematic observables will be given later (Sec. 9).

3. Structure of the physical Hilbert space

Quantum mechanics postulates an isomorphic mapping of observables into the self-adjoint operators on Hilbert space. It is assumed that all physical statements are uniquely given by this mapping, but the converse is not true. As we shall see, there are equivalence classes of mappings which are physically indistinguishable, and we shall make use of this ambiguity to choose the simplest operators.

The original formulation of quantum mechanics by v. NEUMANN and DIRAC postulated a one-to-one correspondence between observables and all self-adjoint operators on Hilbert space. It has become clear since then that there are self-adjoint operators which are not observables, so that the mapping must be *into* and not *onto* the self-adjoint operators [10].

It is convenient to envelop the observables (we drop the distinction between operationally-defined observables and those self-adjoint operators which represent them) in a ring R of operators. If A, B are observables, then $aA + bB$ (a, b complex numbers) and AB are members of the ring.

As a substitute for the old postulate, we assume that the algebra of observables is algebraically closed in the sense that all self-adjoint members of the ring enveloping the observables are observables. We do not need to decide on the inclusion of limits of sums (i. e., topological closure) in the ring. The old postulate (every self-adjoint operator is an observable) evidently implies closure, but closure does not imply the former assumption.

A ring of operators can be decomposed into a direct integral of irreducible subrings. In a finite-dimensional Hilbert space, this decomposition is discrete, i. e., every member A of the ring can be written as

$$A = \sum P_i A P_i, \tag{3.1}$$

where the P_i are a fixed complete set of commuting projection operators with projection subspaces $\mathcal{H}_i$. In this decomposition, the restriction of the ring to one of the subspaces is irreducible, i. e., the only operator of the type $P_i Q P_i$ which commutes with all $P_i A P_i$ $(A \in R)$ is $c P_i$ (c is a complex number). Consider a ring R with a reduced form

$$A = \sum_i \sum_{j=1}^{M_i} P_{ij} A P_{ij} \quad (\text{all } A \in R), \tag{3.2}$$

where all unitarily equivalent terms have the same first index i; i. e., there exist unitary operators U_{il} such that

$$P_{ij} A P_{ij} = U_{il} P_{il} A P_{il} U_{il}^{-1} \quad (j, l \leq M_i, \text{ all } A \in R). \tag{3.3}$$

The decomposition into subspaces $\mathcal{H}_i$ with fixed index i (factors) is unique, while a trivial rotation of the equivalent subspaces is still arbitrary.

In an infinite-dimensional Hilbert space, the decomposition is not necessarily discrete. However, the more exotic types of algebras are of no physical significance.

What mathematical objects represent states, or, more generally, statistical ensembles? This question was raised and answered by v. Neumann [11] under the assumption of a one-to-one mapping of observables on self-adjoint operators, and we have to re-examine it without this restriction.

We ask for a real-valued function Exp on observables A which is:
(a) linear, i. e.,

Exp$(\varrho, aA + bB) = a$ Exp$(\varrho, A) + b$ Exp(ϱ, B) $(a, b$ real numbers$)$;

(b) positive for positive observables $A A^+$, i. e.,

$$\text{Exp}(\varrho, A A^+) \geq 0;$$

(c) normalized, i. e.,

$$\text{Exp}(\varrho, 1) = 1.$$

The first of v. Neumann's results can be taken over because it does not make use of the irreducibility of the ring R. It is the assertion that each function Exp can be implemented by a density operator ϱ so that

$$\text{Exp}(\varrho, A) = \text{Tr} \varrho A. \tag{3.4}$$

However, if ϱ and σ are two density operators, we cannot infer from

$$\text{Exp}(\varrho, A) = \text{Exp}(\sigma, A) \quad (\text{for all } A \in R)$$

that $\varrho = \sigma$, as v. NEUMANN does, since not all self-adjoint operators are in R. It is easy to show that for the decomposition (3.2) of R, the implication of

$$\operatorname{Tr}\varrho A = \operatorname{Tr}\sigma A \quad \text{(all } A \in R) \tag{3.5}$$

is not $\varrho = \sigma$; for the simplest case where the factors have multiplicity one, i. e., $M_i = 1$ for all i, the inference from Eq. (3.5) is only

$$P_i \varrho P_i = P_i \sigma P_i . \tag{3.6}$$

Thus, infinitely many density operators correspond to the same statistical ensemble. This unpleasant one-many relationship between physical objects and their mathematical image can be corrected by the convention: Only those density operators which are members of R will be considered as representatives of physical ensembles:

$$\varrho \in R \quad \text{(1st convention) .} \tag{3.7}$$

Then the choice of the density operator ϱ for a given function $\operatorname{Exp}(\varrho, \ldots)$ is unique, as can be easily verified.

We now have a mapping of physical ensembles into density operators, rather than a one-to-many relation. However, it is still unsatisfactory that not every density operator which commutes with the projections P_i corresponds to an ensemble. Indeed, if two terms in the reduction of R are unitarily equivalent, then, according to our convention, only density operators with this property are admissible representatives of ensembles.

Again, the situation can be simplified by a convention which makes use of the equivalence of different rings R. Compare a ring R [Eq. (3.2)] with a ring R_t which differs from R in that all but one of the equivalent terms have been omitted so that

$$A_t = \sum_i P_{i1} A P_{i1} \quad \text{(all } A_t \in R_t) \tag{3.8}$$

To every density operator $\varrho \in R$, we can associate a density operator

$$\varrho_t = \sum_i \frac{P_{i1} \varrho P_i}{M_i} (\operatorname{Tr}\varrho_t)^{-1} . \tag{3.9}$$

Clearly, it follows that

$$\operatorname{Tr}\varrho A = \operatorname{Tr}\varrho_t A_t \tag{3.10}$$

because the term $\operatorname{Tr}\sum P_{i1} \varrho A P_{i1}$ appears M_i times in $\operatorname{Tr}\varrho A$. Between a ring R and its truncated equivalent R_t and between the corresponding density operators ϱ and ϱ_t there is a one-to-one association such that all expectation values are equal. Hence, we may, without loss of generality, select R_t as the representative of the equivalence class. By convention, we shall deal only with "truncated" rings. In technical language, we consider only rings where factors have multiplicity 1. At the risk of being repetitious we restate that by this convention, no two restrictions $P_i A P_i$ are unitarily equivalent:

$$U P_i A P_i U^{-1} \neq P_j A P_j$$
$$\text{(for some } A \in R, \; i \neq j, \text{ all } U)$$
$$\text{(2nd convention) .} \tag{3.11}$$

With this convention, the situation is simplified because now any density operator ϱ which commutes with all projections P_i represents a physical ensemble, since it belongs to R.

It is now easy to apply v. Neumann's further arguments to each subspace $\mathscr{H}_i$ in proving that pure states are represented by minimal projections (rays) which are entirely in one of subspaces $\mathscr{H}_i$, and that every such ray represents a pure state. Summarizing, we have the following structure of "physical Hilbert space":

It is the union (not the direct sum) of a complete orthogonal set of subspaces $\mathscr{H}_i$. The restriction R_i of the ring of observables R to each subspace $\mathscr{H}_i$ is irreducible, i. e., it commutes only with multiples of the unit operator. Since all restrictions R_i are inequivalent, the only operators that commute with R are constants d_i on each subspace $\mathscr{H}_i$. These commute with each other, i. e., the commutant of R is Abelian.

An ensemble is represented by a density operator ϱ which leaves each subspace $\mathscr{H}_i$ invariant, and every such density operator represents an ensemble. A pure state is represented by a ray which is entirely in one of the subspaces $\mathscr{H}_i$, and every such ray represents a state.

These properties of $\mathscr{H}$ follow from the postulated closure of the ring of observables, and from two conventions of convenience.

It is true that our conclusions are correct only to the same extent as v. Neumann's, i. e. only formally. In fact, there exist "pathological" ensembles which cannot be implemented by density operators on Hilbert space [8]. There are two ways to justify the simple result: either one can postulate that ensembles should also be *probability measures* on observables [9], or one can show that it is physically unnecessary to consider the "pathological" ensembles, because they can be arbitrarily approximated by those implementable by density operators [12].

4. Symmetry operators on Hilbert space

A ray f is the collection of vectors of unit length which differ from each other by phase factors δ, i. e.,

$$f = \{fe^{i\delta}\}, \quad \delta \text{ real}, \quad |f| = 1. \tag{4.1}$$

Two state rays f and g are coherent if $(af + bg)\,(\|af + bg\|)^{-1}$ is contained in a state ray. According to Sec. 3, a ray represents a state if and only if it is contained in the union (not the direct sum) $\bigcup \mathscr{H}_i$ of the subspaces $\mathscr{H}_i$. Two state rays f and g are coherent if and only if they are contained in the same subspace $\mathscr{H}_i$. A subspace $\mathfrak{h}$ is coherent if all rays contained in it are states. Clearly, the $\mathscr{H}_i$ are maximally coherent subspaces, i. e., all rays coherent with any ray in $\mathscr{H}_i$ are in $\mathscr{H}_i$.*

A permutation of states induced by an element L of G is represented by a permutation $f \to U(L)\,f$ which leaves $\bigcup \mathscr{H}_i$ invariant, for Uf must

---

* The maximally coherent subspaces will be denoted by upper case letters $\mathscr{H}$, other subspaces (coherent or incoherent) by lower case letters $\mathfrak{h}$.

be a state. Two successive transformations L_1, L_2 must combine to give the same result as $L_1 L_2$, i. e.,

$$U(L_1)\, U(L_2) = U(L_1 L_2) \,. \tag{4.2}$$

In Sec. 3, we considered symmetries merely as simultaneous permutations of observables and ensembles. In quantum mechanics, both spaces have considerable structure, and experience shows that the symmetry permutations preserve these structures, i. e. they are *automorphisms*. Specifically, in the space of density operators, convex linear combinations are preserved, i. e.

$$a\, \varrho_1 + b\, \varrho_2 \to a\, \varrho_1' + b\, \varrho_2'$$
$$(a + b = 1) \,. \tag{4.3}$$

For a density operator which is the sum of two states represented by one-dimensional projections P_1 and P_2 we have

$$\frac{1}{2}\,(P_1 + P_2) \to \frac{1}{2}\,(P_1' + P_2') \,.$$

Hence, the trace of the square $(P_1 + P_2)^2$ is preserved

$$\mathrm{Tr}\,(P_1 + P_2)^2 = \mathrm{Tr}\,P_1 + \mathrm{Tr}\,P_2 + 2\,\mathrm{Tr}\,P_1 P_2 \,. \tag{4.4}$$

If ψ_1, ψ_2 are two vectors in the rays belonging to P_1 and P_2, the last term of Eq. (4.4) gives

$$2\,\mathrm{Tr}\,P_1 P_2 = 2\,(\psi_2,\, P_1\, \psi_2) = 2\,|(\psi_1,\, \psi_2)|^2 \,. \tag{4.5}$$

Hence, the "distance"

$$d\,(f_1,\, f_2) = 2\,[1 - |(\psi_1,\, \psi_2)|^2] \tag{4.6}$$

is preserved by an automorphism

$$d\,(f_1,\, f_2) = d\,(U f_1,\, U f_2) \,. \tag{4.7}$$

It stands to reason that a small movement (or acceleration) of the apparatus should induce an automorphism little different from the unity. More precisely, we require for physical reasons that the realization of the group G by ray operators $U(L)$ should be continuous.

From Eq. (4.1), one proves (Appendix 1):

T.1. *An operator $U(L)$ maps all rays of a coherent subspace $\mathcal{H}_i$ onto the totality of the rays of a coherent $\mathcal{H}_j$, where possibly $i = j$.*

We consider now the restriction U_i of a ray operator to the maximally coherent $\mathcal{H}_i$. This U_i is a mapping of $\mathcal{H}_i$ onto $\mathcal{H}_j$ which preserves convex combinations.

It can be proved [13] that such a mapping can be implemented by semilinear operators. More precisely, we have:

T.2. *It is possible to choose semilinear* vector operators $\tilde{U}_i$ such that*

$$U_i f = g \quad \text{if} \quad U_i f = g, \quad f \in f, \quad g \in g \,. \tag{4.8}$$

---

* Semilinear means, in the case of Hilbert space, either linear or antilinear. An antilinear operator R is defined by $R\,(a f + b g) = a^* R f + b^* R g$.

Since all rays have unit length, the vector mapping is length-preserving, and $\tilde{U}_i$ is either unitary (if it is linear) or antiunitary, i. e.,

$$(Uf, Ug) = \begin{cases} (f, g) \\ (f, g)^* \, . \end{cases} \qquad (4.9)$$

Instead of the set of operators $\tilde{U}_i(L)$ with domain $\mathscr{H}_i$, we may define a single partial isometry $\tilde{U}(L)$ with domain and range $\bigcup_i \mathscr{H}_i$, whose restriction to $\mathscr{H}_i$ is $\tilde{U}_i$.

Given a set of restrictions $\tilde{U}_i(L)$, clearly any other set

$$\tilde{U}_i(L) = w_i(L) \, \tilde{U}_i(L) \quad (|w_i| = 1)$$

corresponds to the same ray operators $U_i(L)$, and therefore to $U(L)$. If $D(L)$, $D'(L)$ are diagonal operators such that $D(L) \, \phi_i = w_i \, \phi_i (\phi_i \in \mathscr{H}_i; |w_i| = 1)$, then any partial isometry

$$\tilde{U}'(L) = D(L) \, \tilde{U}(L) \, D'(L) \qquad (4.10)$$

is equivalent to $\tilde{U}(L)$ in that it induces the same ray mapping $U(L)$. We may now consider each ray operator $U(L)$ as a class of isometries

$$U(L) = \{D(L) \, \tilde{U}(L) \, D'(L)\} \qquad (4.11)$$

whose elements are obtained by multiplying a fixed (representative) isometry $\tilde{U}(L)$ by all diagonal isometric operators $D(L)$, $D'(L)$.

By Eq. (4.2), the representative isometries will satisfy

$$\tilde{U}(L_1) \, \tilde{U}(L_2) = D(L_1, L_2) \, \tilde{U}(L_1 L_2) \, D'(L_1, L_2) \, . \qquad (4.12)$$

This is the essential mathematical point in which the present study differs from Wigner's which discussed the simpler multiplication table:

$$U(L_1) \, U(L_2) = w(L_1, L_2) \, U(L_1 L_2) \quad (|w| = 1) \, . \qquad (4.13)$$

The complicating diagonal operators are the price which we have to pay for reducing the problem of realization by transformations of a nonlinear space $\bigcup \mathscr{H}_i$ to a realization problem on ordinary linear Hilbert space.

It is convenient to extend the domain and range of the vector isometries $\tilde{U}(L)$ in the natural way. If $\phi_i \in \mathscr{H}_i$, $\phi_j \in \mathscr{H}_j$, we define

$$U(L) \, (\phi_i + \phi_j) = \tilde{U}(L) \, \phi_i + \tilde{U}(L) \, \phi_j \, . \qquad (4.14)$$

By definition, the unitary or antiunitary operators $U(L)$ satisfy the multiplication rule, Eq. (4.12) but only their restriction to $\bigcup \mathscr{H}_i$, viz. $\tilde{U}(L)$, can be used to construct ray operators $U(L)$ which satisfy Eq. (4.2).

It is possible to simplify the basic equation (4.12) so that only one diagonal operator appears, i. e., so that

$$U(L_1) \, U(L_2) = D(L_1, L_2) \, U(L_1 L_2) \, . \qquad (4.15)$$

We give the proof in Appendix 2.

5. The connected subgroup $G_{+\uparrow}$

The postulated continuity of the ray operators requires that the distance

$$d\,[U(L)\,f,\,g]$$

be a continuous function of L with respect to the group topology. By Eq. (4.6) and by the identity

$$\mathrm{Exp}\,(g,\,P_f) = (g,\,P_f g) = |(f,\,g)|^2 \quad (f \in f,\, g \in g)\,, \tag{5.1}$$

the function

$$s^2\,[U(L)\,f,\,g] = |(U(L)\,f,\,g)|^2 \quad (U \in U,\, f \in f,\, g \in g) \tag{5.2}$$

is continuous in L. Consider a complete set of orthonormal vectors g_n^i in $\mathcal{H}_i$. If $f \in \mathcal{H}_i$,

$$T\,(f) = \sum_n s^2(f,\,g_n^i) = 1 \quad (f \in \mathcal{H}_i,\, f \in f)\,; \tag{5.3}$$

but if $f \in \mathcal{H}_j$ $(i \neq j)$, T vanishes. Consider now

$$T\,[U(L)\,f] = \sum_n s^2\,[U(L)\,f,\,g_n^i] \quad (f \in \mathcal{H}_i,\, f \in f,\, U \in U)\,. \tag{5.4}$$

By T.1, the transforms $U(L)\,f$ must be either in $\mathcal{H}_i$ for all L, or in an orthogonal $\mathcal{H}_j$ for all L. Therefore, $T\,(Uf)$ can have only the values 1 and 0. Since the unit operator is one of the elements L of $G_{+\uparrow}$, at least one functional value of T is 1 by Eq. (5.3). Therefore continuity requires that it be 1 for every L. Hence:

T.3. *The ray operators which represent the connected subgroup $G_{+\uparrow}$ map each maximally coherent subspace $\mathcal{H}_i$ onto itself.*

For convenience, we select representative vector operators $U(L)$ which are strongly continuous, i. e., operators for which it is possible to find a neighborhood of L_1 which contains L_2 so that the magnitude

$$\|U(L_2)\,f - U(L_1)\,f\|$$

is as small as one wishes for all L_2 in this neighborhood. This requirement is imposed for mathematical convenience, to be distinguished from the physical necessity for the continuity of the ray operators $U(L)$.

The mathematical problem reduces then to Eq. (4.9) which has been studied exhaustively by WIGNER and others. According to BARGMANN [14] it is possible to choose strongly continuous representative operators $U(L)$ $(L \in G_{+\uparrow})$ which are isomorphic to the local group. That is, within a suitable neighborhood the operators U satisfy

$$U(L_1)\,U(L_2) = U(L_1 L_2) \tag{5.5}$$

without a factor w.

The operators that satisfy Eq. (5.5) locally form (on the whole sheet $G_{+\uparrow}$) a representation not of $G_{+\uparrow}$, but of its cover group $G'_{+\uparrow}$. The representations for which Eq. (5.5) is true everywhere correspond to integral spin, while the simplest choice for half-integral spin is

$$U(L_1)\,U(L_2) = \pm\,U(L_1 L_2)\,. \tag{5.6}$$

We shall choose these simple phases for representatives of $U(L) \in G_{+\uparrow}$ and will verify later that no generality is lost.

The generators of one-parameter subgroups are, by postulate, identified with observables. We adopt the usual physical assumptions which connect the generators to momentum P, energy P_0, center-of-energy K, and angular momentum J [2].

6. The reflection operators

By virtue of the results of Sec. 3, there exist observables which are numerical block-diagonal nondegenerate matrices, i. e., operators whose restriction to the coherent subspaces $\mathcal{H}_i$ are multiples d_i of the unit operator, with $d_i \neq d_j$. Such operators commute with all observables, and a fortiori with kinematic observables. By postulate (Sec. 2), they are left invariant by time reversal. It follows that the subspaces $\mathcal{H}_i$ are left invariant by time reversal. Since by T.3 the same is true for the representatives of $G_{+\uparrow}$, and since by Eq. (4.11) products of operators differ from the operator representing the product by diagonal matrices only, this result can be extended to elements of the form $TL (L \in G_{+\uparrow})$, i. e., to the sheet $G_{-\downarrow}$.

T.4. *For elements L of the sheet $G_{-\downarrow}$ of G, the operators $U(L)$ leave each space $\mathcal{H}_i$ invariant.* Thus, the only operators that may permute subspaces $\mathcal{H}_i$ belong to the sheets $G_{+\downarrow}$ and $G_{-\uparrow}$. These can be obtained by multiplying the representatives of the orthochronous subgroup $G_{+\uparrow}$ by that of the space inversion operator I and space-time inversion IT. Therefore, for the purpose of discussing incoherent subspaces connected by the Poincaré group, we may restrict our attention to the operators $U(I)$, $U(IT)$, and $U(L')$ such that L' is in $G_{+\uparrow}$.

If $U(I)$ maps a ray f in $\mathcal{H}_i$ onto g in $\mathcal{H}_j$, then, by Eq. (4.2), a second application of $U(I)$ must return the ray into $\mathcal{H}_i$, since

$$I^2 = 1 . \tag{6.1}$$

Clearly, the same argument holds for $U(IT)$. Thus we have:

T.5. *At most, two mutually incoherent subspaces $\mathcal{H}_i$ are connected by symmetry operators.* The ray operator $U(I)$ may map a coherent subspace $\mathcal{H}_i$ either onto itself or onto another coherent subspace. In the former case, incoherence plays no role and we have the case exhaustively discussed in the literature. In the second case, only the two connected subspaces $\mathcal{H}_1$ and $\mathcal{H}_2$ need to be considered. Then, in the natural representation in which diagonal matrices leave each maximally coherent subspace $\mathcal{H}_i$ invariant, a representative vector operator $U(I)$ may be written as

$$U(I) = \begin{pmatrix} 0 & \alpha \\ \beta & 0 \end{pmatrix} , \tag{6.2}$$

where α and β are unitary. The general form of $U(L)$ $(L \in G_{+\uparrow})$ is

$$U(L) = \begin{pmatrix} u_1(L) & 0 \\ 0 & u_2(L) \end{pmatrix} . \tag{6.3}$$

From Eq. (4.11) and the group multiplication table, one finds after some analysis (Appendix 3) that

$$
\begin{pmatrix} \alpha\, u_2(L)\, \alpha^+ & 0 \\ 0 & \alpha^+ u_1(L)\, \alpha \end{pmatrix} = \begin{pmatrix} u_1(ILI) & 0 \\ 0 & u_2(ILI) \end{pmatrix}. \tag{6.4}
$$

If there exists a unitary operator β such that

$$
\beta\, u_1(ILI)\, \beta^+ = u_1(L) \quad (L \in G_{+\uparrow})\,, \tag{6.5}
$$

then the representation u_1 is equivalent to u_2.

7. Irreducible realizations

A group realization by ray operators is irreducible if no proper subspace is left invariant by *all* representative ray operators. If a space $\mathfrak{h}_1 \oplus \mathfrak{h}_2$, containing two separately coherent but mutually incoherent subspaces $\mathfrak{h}_1$ and $\mathfrak{h}_2$, is to be an irreducible realization space, then at least the corresponding projective representations of $G_{+\uparrow}$, i. e., u_1 and u_2 in Eq. (6.3) must be irreducible.

Irreducible representations are intrinsically characterized by the numerical values assigned to a complete set of invariants. If the (outer) automorphism $L \to ILI$ leaves all invariants unchanged, then the representation $u(ILI)$ is equivalent to $u(L)$. A full list of invariants for $G_{+\uparrow}$ is given by SHIROKOV [15]. For the representations of physical importance, they are

$$
m^2 = P_\mu P_\mu\,; \quad s(s+1) = m^{-2}\, \Gamma_\sigma^2\,;
$$

$$
\operatorname{sign} P_0\,;
$$

$$
\delta(P_\mu P_\mu)\, \frac{\boldsymbol{J} \cdot \boldsymbol{p}}{|\boldsymbol{p}|}\,. \tag{7.1}
$$

The first three invariants are unchanged by space inversion. The last one vanishes except for representations with $P_\mu P_\mu = -m^2 = 0$; but for these massless representations, this invariant (the helicity) changes sign under space inversion.

Thus we have the alternative: If two irreducible representation spaces of $G_{+\uparrow}$ or $G'_{+\uparrow}$ are mutually incoherent and connected by space inversion, then they are both either of nonvanishing mass ($m \neq 0$) and mutually equivalent, or they are of zero mass and inequivalent.

We have investigated the structure of ray realizations on the assumption that the two coherent subspaces were mutually incoherent. To show that they must indeed be incoherent in some cases, we need an algebraic theorem:

T.6. *In an n-dimensional vector space $(n > 1)$ there exists no anti-unitary operator that commutes with all linear Hermitean operators.*

11*

Proof (by *reductio ad absurdum*): Consider two Hermitean operators A, B that do not commute. For $n > 1$, such operators always exist. The commutator $[A, B]$ is anti-Hermitean; therefore, it may be written as iC, where C is Hermitean. Consider the equation

$$[A, B] = iC , \tag{7.2}$$

and multiply from the left by an antiunitary operator R, and from the right by its inverse R^{-1}. By assumption, R commutes with A, B, and C. Hence, the left-hand side remains unchanged, while the right-hand side changes sign. Thus, we have

$$[A, B] = iC = -iC , \tag{7.3}$$

i. e., $C = 0$ contrary to assumption.

Returning to the incoherent subspaces, we notice that for $m \neq 0$ the two representations u_1 and u_2 are equivalent and therefore may be assumed to be equal without restriction of generality. Then, in the space $\mathfrak{h}_1 \oplus \mathfrak{h}_2$, all operators of the form $M \otimes 1$ (where M is a two-dimensional numerical matrix) will commute with $U(L) = 1 \otimes u(L)$ $(L \in G_{+\uparrow})$, and therefore with all generators. For a single particle, the kinematic observables are the generators of $G_{+\uparrow}$, p, p_0, J, and K, and their linear combinations (the elements of the Lie algebra Γ). By our postulate in Sec. 2, all observables which commute with them, and hence with all $U(L)$ $(L \in G_{+\uparrow})$, must be time-reversal invariant.

All Hermitean matrices M that are observables should then commute with $U(T)$. Theorem 6 asserts, however, that this is impossible, so that not all Hermitean operators in the space $\mathfrak{h}_1 \oplus \mathfrak{h}_2$ can be observables. It follows that the assumed incoherence between $\mathfrak{h}_1$ and $\mathfrak{h}_2$ is, in fact, a necessity for irreducible representations (reps) with $m \neq 0$.

However, this proof does not cover the case $m = 0$, since then $u_1 \neq u_2$ in any realization, and therefore only diagonal matrices $M \otimes 1$ commute with $U(L)$ $(L \in G_{+\uparrow})$. By virtue of the principle of maximal coherence, we must deny the existence of incoherence if there is no reason for it. We have, therefore:

T.7. *Two rep spaces of $G_{+\uparrow}$ or $G'_{+\uparrow}$, connected by space inversion, are mutually incoherent if and only if they are equivalent, i. e., if and only if the mass is different from zero.*

We now describe the irreducible representatives (Appendix 4). Let $U_0(L)$ be a projective rep $(m \neq 0)$ of the type described by Wigner, i. e., on a rep space of $G'_{+\uparrow}$; for these representations, the number of linearly independent states with given p (the multiplicity) is $2s + 1$. Our representation space consists of two such spaces, attached incoherently.

The irreducible realizations for $m \neq 0$ may be written as the direct product of a two-by-two matrix and $U_0(L)$ in the natural representation in which $\binom{\phi}{0}$, $\binom{0}{\phi}$ are states but their linear combinations with non-vanishing coefficients are not. The general form, with arbitrary diagonal matrices D, is

$$U(L) = D(L) \begin{pmatrix} 1 & 0 \\ 0 & 1 \end{pmatrix} \otimes U_0(L) \quad (L \in G_{+\uparrow}),$$

$$U(I) = D(I) \begin{pmatrix} 0 & 1 \\ 1 & 0 \end{pmatrix} \otimes U_0(I); \quad U(T) = D(T) \begin{pmatrix} 1 & 0 \\ 0 & 1 \end{pmatrix} \otimes U_0(T),$$

$$U(IT) = D(IT) \begin{pmatrix} 0 & 1 \\ 1 & 0 \end{pmatrix} \otimes U_0(IT). \tag{7.4}$$

It may be useful to compare these operators with the corresponding projective representations of the Poincaré group, with multiplicity $2(2s+1)$, on a coherent space [7]. For the irreducibility of these, it is decisive that $D(I)$ and $D(IT)$ should be of the form $\begin{pmatrix} 1 & 0 \\ 0 & -1 \end{pmatrix}$, because otherwise a unitary transformation of the representation space $\mathfrak{h}_1 \oplus \mathfrak{h}_2$ could reduce all operators simultaneously. In our case the choice of the matrices $D(L)$ has no importance at this point; the irreducibility is due to the incoherence of the subspaces $\mathfrak{h}_1$ and $\mathfrak{h}_2$ which, in turn, is caused by the physical nature of the automorphism of observables induced by the time-reversal element T.

From a few general assumptions, we have derived the following results:

(1) The existence of rigorous mass doublets, i. e., the existence of particles which have $2(2s+1)$ linearly independent states for given mass and momentum. If one accepts the principle: "Accidents don't happen," the statement is that only mass singlets and doublets can exist. If it is remembered that this theory treats only *rigorous* symmetries, the existence of an approximate mass degeneracy between π^0 and the $\pi^\pm$ doublet does not contradict this result.

(2) The symmetry operator which represents space inversion changes particles into antiparticles. It is essentially the well-known operator PC.

(3) Particles and antiparticles are mutually incoherent.

It is true that the geometric explanation of the doublets is not quite as compelling as the explanation of rotational degeneracy in spectra since it is not possible by a physical motion of the producing equipment to change particles into antiparticles. This is due to the disconnected nature of the reflection operations.

8. Assumptions about many-particle states

In the distant past and future, physical systems are decomposed into noninteracting corpuscles. The transformation $U(L)$ induces a transformation of each individual corpuscle in accordance with its one-corpuscle transformation properties [1]. To obtain a formal equivalent of this intuitive statement, we introduce asymptotic creation operators

$a_i^+(f)$ which are linear functions of the vectors of an irreducible subspace. If α, β are complex numbers and f, g are elements of the irreducible subspace, then

$$a_i^+(\alpha f + \beta g) = \alpha a_i^+(f) + \beta a_i^+(g) . \tag{8.1}$$

Given a set of representative vector operators $U(L)$, we postulate that an element L of the orthochronous Poincaré group $(G_{+\uparrow} + G_{-\uparrow})$ should induce an automorphism of the creation operators

$$U(L)\, a^+(f)\, U^{-1}(L) = \eta(N, L)\, a^+[U(L)\, f] . \tag{8.2}$$

The phase factor η, which may be a function of the number operator N, is necessary for generality if the transformation properties of many-particle states are described only by our physical statement.

It will become clear that the principle of maximal coherence requires η to be independent of N. Since $U(L)\, f$ is determined only up to a phase factor depending on L, there is no further loss of generality if we set $\eta = 1$. The definition of the operators a^+ is completed by the usual commutation and anticommutation relations and by the requirement that their adjoints annul the vacuum state Ψ_0. For the elements $TL\,(L \in G_{+\uparrow}$ or $G_{-\uparrow})$ of the sheets $G_{+\downarrow}$ and $G_{-\downarrow}$, we introduce, following Haag [1], antiunitary operators $U^{\text{out}}_{\text{in}}(TL)$ which differ from the true representers $U(TL)$ through multiplication by the scattering operator S, i. e.,

$$\begin{aligned} U(TL)\, S &= U^{\text{out}}(TL) \\ S\, U(TL) &= U^{\text{in}}(TL) . \end{aligned} \quad (L \in G_{+\uparrow} \text{ or } G_{-\uparrow}) \tag{8.3}$$

They are defined (up to multiplicative factors $e^{i\delta}$) by

$$U^{\text{out}}_{\text{in}}(TL)\, a^{+\,\text{out}}_{\text{in}}(f)\, U^{\text{out}}_{\text{in}}(TL) = a^{+\,\text{out}}_{\text{in}}\{U(TL)\, f\} \quad (L \in G_{+\uparrow} \text{ or } G_{-\uparrow}). \tag{8.4}$$

The operator U on the right-hand side has no superscript because the in and out states coincide for one-corpuscle states.

We consider the operational meaning of incoherence. What is the experimental meaning of the statement that a state Ψ is incoherent with the vacuum? As an example, one may think of Ψ as a state of a compound particle of spin $\frac{1}{2}$. One implication of incoherence is the prediction that this compound state cannot disintegrate into an asymptotic state that is coherent with the vacuum, e. g., a state of any number of photons. (For the moment, we accept without proof that the n-photon state is coherent with the vacuum. For proof see Sec. 11.) The verification of such a "global" statement would require a search for particles other than photons in the whole universe; the presence of a spin-$\frac{1}{2}$ particle in the final state, no matter how distant from the original particle, would belie the result. Clearly, such a universal search is experimentally impossible. Hence, if the incoherence statement is to be experimentally verifiable at all, it must be strengthened so that it refers only to a finite space-time volume (principle of local verifiability) [16]*. In this spirit, we postulate that incoherence between Ψ and Ψ_0 shall imply incoherence between any two states that are *locally* indistinguishable from Ψ and Ψ_0, respectively. To put this idea into formal language, consider a Schrödinger state $A^+\Psi_0$, where A^+ is a creation operator for particles that are localized in some finite volume V. Consider also states $\Psi(V')$ which differ from Ψ_0 only outside a large but finite volume V', which includes V. The principle of verifiability requires that

$$(A^+\Psi_0 \text{ incoh } \Psi_0) \quad \text{implies} \quad [A^+(V)\,\Psi(V') \text{ incoh } \Psi(V')]. \quad (8.5)$$

According to T.3, a state Ψ is coherent with $U(L)\,\Psi$ if $L \in G_{+\uparrow}$. Therefore, if L is a time translation, the incoherence between two states persists if each of them is translated in time. Thus

$$(A^+\Psi_0 \text{ incoh } \Psi_0) \quad \text{implies} \quad [A^+(V, t)\,\Psi(V', t) \text{ incoh } \Psi(V', t)] \quad (8.6)$$

for all values t. We have now, on the right-hand side, a huge collection of operators and states which do not have the particular local character of the original ones. It is reasonable to assume that the statement can be extended to all states and creation operators of the given type. We obtain then the generalized form of Eq. (8.5):

$$(A^+\Psi_0 \text{ incoh } \Psi_0) \rightarrow (A^+\Psi \text{ incoh } \Psi). \quad (8.7)$$

9. Observables on many-particle states; automorphism induced by T

In Sec. 7, we have defined a set Γ of kinematic observables for one-particle states. If all particles were distinguishable, we could define one-particle operators P_i, J_i, etc. on many-particle states and obtain a complete algebra of kinematic observables on these states. Because of the indistinguishability of particles, we must proceed in a different way.

Let $f_l^{(i)}$ be a one-particle eigenstate of an observable B in the ith irreducible subspace. Consider the operator

$$a^+(f_l^{(i)})\,a(f_l^{(i)}) = n(f_l^{(i)}). \quad (9.1)$$

Equation (9.1) is meant to hold for either in- or out operators. In the following, we shall omit superscripts whenever it is clear that all creation and destruction operators may be given simultaneously either the "out" or the "in" superscripts. Let ν_l^i be the number of degenerate eigenstates $f_{l,\nu}$ of B that belong to the eigenvalue l in the ith irreducible subspace.

---

* A similar situation arises with respect to many fundamental principles of physics. The principle of energy conservation requires only that the energy of the universe should be the same at two times — an evidently unverifiable statement. The principle becomes directly verifiable because it is combined with statements concerning the local transmission of energy (the local, or at least quasi-local, energy-momentum conservation).

Then the number of (i)-corpuscles having eigenvalues l of an observable B is

$$n_i(B, l) = \sum_{\nu=1}^{\nu_l^i} n(f_{l,\nu}^{(i)}) .\qquad(9.2)$$

It is convenient to define a quantity $n(P_l^i)$, the number of corpuscles whose one-particle eigenvalues of B are not larger than l. It is given by the expression

$$n(P_l^i) = \sum_{\lambda=-\infty}^{l} n_i(B, \lambda) ,\qquad(9.3)$$

where P_l^i can be defined as a projection operator onto the lth subspace of the ith irreducible subspace, i. e., onto that subspace which belongs to the eigenvalue l of an observable B. The relation $P_l^i \to n(P_l^i)$ defines a mapping from one-particle to many-particle observables which is well defined even for observables with continuous spectra [16].

Consider now the kinematic observables of the set Γ. If P_l^i are all the projection operators of the algebra Γ, then the mapping $P_l^i \to n(P_l^i)$ induces an algebra $\tilde{\Gamma}$ which may be considered as the algebra of kinematic observables for many-body states.

Clearly, the set $\tilde{\Gamma}$ is not a complete set of observables for it commutes, e. g., with such operators as the number operator

$$\sum_{i,l} n(f_l^{(i)}) = N ,\qquad(9.4)$$

where the summation extends over all irreducible subspaces (i) and a complete set of states $f_l^{(i)}$ in each of them.

Not even the projection $\tilde{\Gamma}$ into a subspace of definite numbers N_i of particles and $\bar{N}_i$ of antiparticles of kind i is complete. Indeed, the observables $n(P_l^i)$ do not distinguish between particles and antiparticles of the ith corpuscle, and there exist nonkinematical observables which commute with $\tilde{\Gamma}$ and whose restriction to a subspace of given N_i, $\bar{N}_i$ is nontrivial.

Consider observables A that commute with all observables of $\tilde{\Gamma}$. They are, physically, those observables that are needed to characterize a many-body state in addition to, and independently of, its specification by the number of particles having given values of kinematic one-particle observables, e. g., charge or baryon number.

One may be tempted to interpret the physical statement of Sec. 2 by requiring that observables of the class A should remain invariant under the isomorphism induced by time-reversal. We must remember, however, that the observables have been defined with respect to asymptotic in or out states that are carried into each other by time-reversal.

Time-reversal induces a mapping of states. According to Sec. 8, this mapping may be considered in two steps: first it maps an "in" state into another "in" state through the antiunitary operation $U^{in}(T)$, and then it maps the resulting "in" state into the corresponding "out" state by the unitary operation S^{-1}.

We are concerned here only with the first step. Empirical evidence suggests that the time-reversed state does not differ from the original state in such variables as charge and baryon number, i. e., the first step of the time-reversal operation leaves observables of the class A invariant. Formally, we assert as a physical assumption that, if A is an observable, and

$$[A, \overset{\text{in}}{\tilde{\Gamma}^{\text{out}}}] = 0 , \tag{9.5}$$

then

$$[A, \overset{\text{in}}{U^{\text{out}}}(T)] = 0 . \tag{9.6}$$

10. Incoherences between many-body states

Consider now an irreducible subspace $\mathfrak{h}_i$ with two mutually incoherent subspaces $\mathfrak{h}^i_+$ and $\mathfrak{h}^i_-$. By the construction described in Sec. 8, we can obtain many-particle states of n particles and $\bar{n}$ antiparticles of the ith corpuscle. Together with this subspace $\mathfrak{h}_{(n, \bar{n})}$ we consider the subspace of $\bar{n}$ particles and n antiparticles of corpuscle i, viz. $\mathfrak{h}_{(\bar{n}, n)}$. The restriction of $\tilde{\Gamma}$ to $\mathfrak{h}(n, \bar{n})$, viz. $\tilde{\Gamma}(n, \bar{n})$, will be shown to be equal to the restriction $\tilde{\Gamma}(\bar{n}, n)$ of $\tilde{\Gamma}$ to $\mathfrak{h}(\bar{n}, n)$. By the construction of the irreducible incoherent subspaces, the restrictions of Γ to the two mutually incoherent subspaces $\mathfrak{h}_{(\pm)}$ are equal. On the other hand, the construction (9.2) of $n(\Gamma, l)$ and therefore of $n(P_l)$ includes a summation over both subspaces $\mathfrak{h}_{(\pm)}$ and therefore makes $\tilde{\Gamma}$ invariant with respect to an interchange of particles and antiparticles. This proves that $\tilde{\Gamma}(n, \bar{n}) = \tilde{\Gamma}(\bar{n}, n)$.

We now apply the argument given in Sec. 7. In the space $\mathfrak{h}_{(n, \bar{n})_i} \oplus \oplus \mathfrak{h}_{(\bar{n}, n)_i}$, there are Hermitean operators A which commute with the restriction of $\tilde{\Gamma}$ to this subspace. They include all operators of the form $M \otimes \tilde{\Gamma}$, where M is a numerical 2×2 matrix. If they are to be observables, then, by Eq. (9.6), they must commute with $\overset{\text{in}}{U^{\text{out}}}(T)$. This, however, is impossible by T.6 of Sec. 7. We conclude that not all Hermitean operators in $\mathfrak{h}_{n, \bar{n}} \oplus \mathfrak{h}_{\bar{n}, n}$ are observables, and that, therefore, $\mathfrak{h}_{n, \bar{n}}$ and $\mathfrak{h}_{\bar{n}, n}$ are mutually incoherent. This argument assumes, of course, that $\mathfrak{h}_{n, \bar{n}}$ and $\mathfrak{h}_{\bar{n}, n}$ are distinct subspaces, and fails if $n = \bar{n}$. Hence

T.8. *States of n particles and $\bar{n}$ antiparticles $(n \neq \bar{n})$ of a given kind of corpuscle are incoherent with states of $\bar{n}$ particles and n antiparticles of the same kind of corpuscle.*

We digress to make a few remarks on the logical nature of the coherence relation. We want to show that it is, in physical Hilbert space, an equivalence relation. Without the specific restrictions on the structure $\mathscr{H}$, we could not infer from the coherence between the two pairs of vectors (f, g) and (f, h) that g and h are coherent.

However, the physical Hilbert space is a sum of mutually orthogonal maximally coherent subspaces $\mathscr{H}_i$ which are mutually incoherent. It follows that if a ray f is in $\mathscr{H}_1$ and also in $\mathscr{H}_2$, then $\mathscr{H}_1 = \mathscr{H}_2$.

If f and g are coherent, they must be in one subspace $\mathscr{H}_1$; if g and h are coherent, they must both belong to $\mathscr{H}_2$; since g is both in $\mathscr{H}_1$ and $\mathscr{H}_2$, $\mathscr{H}_1 = \mathscr{H}_2$; since f and h are also in $\mathscr{H}_2$, they are mutually coherent. We therefore have:

T.9. *Coherence is an equivalence relation with transitivity;* (f coh g) *and* (g coh h) *imply* (f coh h).

Consider now a subspace of $\mathscr{H}$ which is the direct sum of the three orthogonal subspaces $\mathfrak{h}_{n,\bar{n}}$, $\mathfrak{h}_{\bar{n},n}$, and the vacuum Ψ_0. We wish to prove that the three subspaces are mutually incoherent. The proof is by *reductio ad absurdum*. If $\mathfrak{h}_{n,\bar{n}}$ were coherent with Ψ_0, then a vector $\Psi_{n,\bar{n}} + \Psi_0$ would be a state; an application of the operator $U(I)$ to this state would produce another state of the form $\Psi_{\bar{n}n} + \Psi_0$. Therefore, Ψ_0 would be coherent with $\mathfrak{h}_{\bar{n},n}$ as well as with $\mathfrak{h}_{n,\bar{n}}$. But by T.9 this would imply that $\mathfrak{h}_{n,\bar{n}}$ is coherent with $\mathfrak{h}_{\bar{n},n}$, contrary to T.8.

It follows that all many-body subspaces $\mathfrak{h}_{n,\bar{n}}(n \neq \bar{n})$ are incoherent with Ψ_0, but we do not yet know their mutual coherence relations. For this purpose we use the principle of local verifiability. According to the previous result and Eq. 8.1, a creation operator $A^+_{\nu,\bar{\nu}}$ which creates ν particles and $\bar{\nu}$ antiparticles ($\nu \neq \bar{\nu}$) creates a state incoherent with Ψ when it acts on Ψ.

If Ψ is chosen to be a state with n particles and $\bar{n}$ antiparticles, we have the result that the subspace $\mathfrak{h}_{n+\nu,\bar{n}+\bar{\nu}}$ is incoherent with $\mathfrak{h}_{n,\bar{n}}$-provided $\nu \neq \bar{\nu}$. It is clear then that coherence between two such subspaces $\mathfrak{h}_{\bar{n},n}$ and $\mathfrak{h}_{m,\bar{m}}$ depends only on the differences $n - \bar{n}$ and $m - \bar{m}$; if these differences are unequal, the subspaces are mutually incoherent.

We have derived from general principles the "additivity of generalized charge" (or rather, one half of it). If we define $n - \bar{n}$ as the generalized charge of a state with n particles and $\bar{n}$ antiparticles of a corpuscle, then the result reads: Many-body states are mutually incoherent if they have different generalized charge.

11. Composition of group realizations by ray operators

We have constructed realizations of G by ray operators $u(L)$ on some subspaces $\mathfrak{h}_{n,\bar{n}}$ with definite particle numbers. In the following sections, we shall try to extend these ray operators so that the extensions still form a realization. According to the principle of maximal coherence, rays will be considered as states if the ray operators extended to them still form a realization of the group.

By selecting suitable representative vector operators $u_i(L)$ for each subspace $\mathfrak{h}_i$, one can try to define vector operators $U(L)$ on $\Sigma \oplus \mathfrak{h}_i$ which satisfy Eq. (4.8) and therefore generate a realization on $\Sigma \oplus \mathfrak{h}_i$.

Consider two mutually orthogonal coherent subspaces $\mathfrak{h}_1$ and $\mathfrak{h}_2$ which are invariant under group transformations and on which two realizations $u_1(L)$ and $u_2(L)$ are defined. If the two subspaces are coherent with each other, there exists a realization of the group by ray operators $U(L)$ on $\mathfrak{h}_1 \oplus \mathfrak{h}_2$ such that the restriction of U is u_1 and u_2 on $\mathfrak{h}_1$ and on $\mathfrak{h}_2$, respectively. It must then be possible to choose representative semilinear operators $u_1(L)$, $u_2(L)$ such that

$$U(L) = \{v(L) \, [u_1(L) \oplus u_2(L)]\} \quad (|v| = 1) . \tag{11.1}$$

By Eq. (4.11),

$$u_i(L_\alpha) \, u_i(L_\beta) = w_i(L_\alpha, L_\beta) \, u_i(L_\alpha L_\alpha) \quad (i = 1, 2; \, |w_i| = 1) \tag{11.2}$$

and

$$U(L_\alpha) \, U(L_\beta) = W(L_\alpha, L_\beta) \, U(L_\alpha L_\beta) \quad (|W| = 1) , \tag{11.3}$$

where $U(L)$ is a member of the class defined by Eq. (11.1). Compatibility of Eqs. (11.1)−(11.3) requires that

$$w_1(L_\alpha, L_\beta) = w_2(L_\alpha, L_\beta) \tag{11.4}$$

for all pairs L_α, L_β. The choice is possible if u_1 and u_2 belong to the same equivalence class (as defined by BARGMANN [14]) of solutions of Eq. (4.9). It is clear at once that two ray representations of $G_{+\uparrow}$ cannot be coherent if one corresponds to integral and the other to semi-integral spin — because the phase factors are equivalent to 1 for the former, but are at best equivalent to ± 1 for the latter [Eq. (5.6)].

We consider next the case for which the domains of the ray operators u_1, u_2 are not linearly closed spaces. According to T.5 we have to consider at most two coherent subspaces $\mathfrak{h}^i_+$, $\mathfrak{h}^i_-$ $(i = 1, 2)$ which are mutually incoherent for fixed i.

If one of the representation spaces is coherent and the other incoherent, then the answer to our question of extended coherence is simple: The two representation spaces are mutually incoherent. This follows from T.9 by the same consideration (Sec. 10) that excludes possible coherence between Ψ_0, $\mathfrak{h}_{n,\bar{n}}$, and $\mathfrak{h}_{\bar{n},n}$.

We are left with the case in which both domains consist of two mutually incoherent subspaces.

By the same reasoning as used above, we may exclude a coherence between three of the coherent subspaces, e. g., $\mathfrak{h}^1_+$, $\mathfrak{h}^2_+$, and $\mathfrak{h}^2_-$. The largest possible coherence is between pairs of coherent subspaces, e. g.,

$$\begin{align} &(\mathfrak{h}^1_+ \text{ coh } \mathfrak{h}^2_+) \\ &(\mathfrak{h}^1_- \text{ coh } \mathfrak{h}^2_-) . \end{align} \tag{11.5}$$

The condition for this coherence is again, as is easily seen, that the phase factors w_1 and w_2 can be chosen to be equal.

In composing operators defined on subspaces we will consider first the system comprising an irreducible subspace of corpuscles of kind α, the vacuum, and many-body states with α corpuscles. We shall choose phase factors for the irreducible subspace and the vacuum so as to maximize the coherence of the set of subspaces considered.

This requirement will impose some constraints on the choice of the phase factors. The remaining freedom will be used to establish maximal coherence between different sets of corpuscle states of different kinds (α, β, etc.) as well as between subspaces with more than one kind of corpuscle.

We can, without loss of generality, make a choice of phase factors for one coherent subspace, viz. the vacuum Ψ_0. As usual, we set

$$U(L)\,\Psi_0 = \Psi_0\,, \tag{11.6}$$

i. e., all phase factors are 1.

We show first that the usual phase assignments for particles without antiparticles (i. e., described by coherent irreducible subspaces) agree with our requirement of maximal coherence. For particles with integral spin, a proper corepresentation of $G(w=1)$ is a member of the equivalence class, and by Eq. (11.4) this is the one we have to choose for coherence with the vacuum. Furthermore, the operator $U(L)$ that Eq. (8.2) (with $\eta = 1$) assigns to each many-particle state is automatically one that forms a proper corepresentation of G. For particles with half-integral spin and no antiparticles (if they exist), coherence with the vacuum cannot be secured since the simplest phase factors are still ± 1. The maximal coherence is obtained precisely by this choice because the phase factors in subspaces with n particles will be $(\pm 1)^{2n}$, i. e., there is coherence between the vacuum and states of even n and states with odd n are coherent with each other. These are, indeed, the conventional assignments. We have shown them to be not only convenient but necessary, once the choice (11.6) is made.

We now consider the particles that have antiparticles. For the representatives of the connected subgroup $G_{+\,\uparrow}$, we make the same choice as above, i. e., $w = 1$ for integral spin and $w = \pm 1$ for semi-integral spin. Clearly, this choice will be the most favorable one for coherence. In fact, this choice was anticipated in Sec. 5.

According to Sec. 10, many-corpuscle states $(n, \bar{n})$ are incoherent with the vacuum unless $n = \bar{n}$. If it is possible to choose 2×2 matrices $D(I)$, $D(IT)$, and $D(T)$ in Eq. (7.4) such that states with $n = \bar{n}$ (constructed by applying products of operators $a^+(f)$ to the vacuum) form a basis for a proper group corepresentation, then these states are coherent with the vacuum.

If we designate the two diagonal elements of the matrices $D(L)$ in Eq. (7.4) by $d_{1,2}(L)$, then Eq. (8.2) applied to a two-corpuscle state with $n = \bar{n} = 1$ for the case in which $L = I$, takes the form

$$
\begin{aligned}
U(I)\, & a^+ [f_+(\boldsymbol{p}, s)]\, a^+ [f_-(\boldsymbol{p}, s)]\, \Psi_0 \\
&= U(I)\, a^+ [f_+(\boldsymbol{p}, s)]\, U^{-1}(I)\, U(I)\, a^+ [f_-(\boldsymbol{p}, s)]\, U^{-1}(I)\, U(I)\, \Psi_0 \\
&= d_1(I)\, a^+ [U_0(I)\, f_-(\boldsymbol{p}, s)]\, d_2(I)\, a^+ [U_0(I)\, f_+(\boldsymbol{p}, s)]\, \Psi_0 \\
&= \pm\, d_1(I)\, d_2(I)\, a^+ [f_+(-\boldsymbol{p}, s)]\, a^+ [f_-(-\boldsymbol{p}, s)]\, \Psi_0\,.
\end{aligned} \tag{11.7}
$$

The $+(-)$ sign applies to Bose (Fermi) statistics. The omission of the operator $U(I)$ in front of Ψ_0 is due to the convention stated in Eq. (11.6).

The requirement that two-particle states be coherent with Ψ_0. mplies that the group G is homomorphic onto the restriction of the operators U^{in} to this two-body space. For instance, by Eq. (11.7) the equation

$$[U(I)]^2 = 1$$

should imply

$$
\begin{aligned}
[d_1(I)\, d_2(I)]^2\, & a^+ [f_+(\boldsymbol{p}, s)]\, a^+ [f_-(\boldsymbol{p}, s)]\, \Psi_0 \\
&= a^+ [f_+(\boldsymbol{p}, s)]\, a^+ [f_-(\boldsymbol{p}, s)]\, \Psi_0\,,
\end{aligned} \tag{11.8}
$$

i. e.,

$$d_1(I) = \pm d_2^{-1}(I) . \tag{11.9}$$

Hence, the most general matrix $D(I)$ in Eq. (7.4) is restricted to be of the form

$$D(I) = \begin{pmatrix} d(I) & 0 \\ 0 & \pm d^{-1}(I) \end{pmatrix} . \tag{11.10}$$

On applying this procedure to the multiplication table of the subgroup $(I, T, IT, 1)$, one finds only one more restriction on the values of the constants $d_{1,2}(L)$, viz.

$$d_1(T) \, d_2(T) \, d_1^{-1}(IT) \, d_2^{-1}(IT) = [U(I)]^2 . \tag{11.11}$$

Since it clearly is possible to meet these requirements, the principle of maximal coherence asserts that the two-corpuscle state with $n = \bar{n} = 1$ is indeed coherent with the vacuum.

Application of the same method to states with equal numbers of particles and antiparticles shows that the same requirements are sufficient to make all states with $n = \bar{n}$ coherent with the vacuum. Finally, one can see that the phase factors for a general state $(n, \bar{n})$ depend only on the difference between the number of creation operators of particles and of antiparticles, so that, quite generally, states with equal difference $n - \bar{n}$ are mutually coherent. Thus, all coherences that were not excluded by Sec. 6 are allowed and, we expect, are realized in nature. Thus we may generalize: *For many-body states of a given corpuscle, states are mutually incoherent if and only if they have different generalized charge $n - \bar{n}$.*

12. Types of doublets

How many different groups of ray operators correspond to different choices of the allowed phase factors $d(I)$ and $d_{1,2}(T)$, $d_{1,2}(IT)$, subject to the constraint (11.11)? Let us remember that the only observable entities are the ray operators $U(L)$ which, at present, we have defined on the union $\mathbf{U}_\alpha$ of all coherent subspaces of equal generalized charge for a given corpuscle α, i. e., on

$$\mathbf{U}_\alpha = \overset{\infty}{\underset{\nu=0}{\mathbf{U}}} \sum_m \oplus \, \mathfrak{h}_{m,m+\nu} .$$

The results of Sec. 11 show that a simple set of irreducible vector operators which can be chosen without loss of generality is

$$\begin{aligned}
U(L) &= 1 \otimes U_0(L) \quad (L \in G_{+\dagger}) \\
U(I) &= \begin{pmatrix} 0 & 1 \\ \pm 1 & 0 \end{pmatrix} \otimes U_0(I) \\
U(T) &= \begin{pmatrix} 1 & 0 \\ 0 & 1 \end{pmatrix} \otimes U_0(T) \\
U(IT) &= \begin{pmatrix} 0 & \pm z \\ z^* & 0 \end{pmatrix} \otimes U_0(IT) \quad (|z| = 1) .
\end{aligned} \tag{12.1}$$

The arbitrary complex number z has no influence on the ray operators, and could be chosen equal to 1 but, for a reason to be given presently, we prefer to keep it general.

There remains the significant choice between the two signs in $U(I)$ and $U(IT)$. This has observable consequences for many-body spaces. Consider a subspace $\mathfrak{h}_{1,1}$, i. e., a subspace of one particle and one antiparticle. The choice in $U(I)$ corresponds to the distinction between the cases in which particle and antiparticle have equal or opposite coparity. The first is believed to be true for pions, the second is certainly true for electrons; the pseudoscalar decay pattern of the positronium ground state is conclusive evidence. It would be pretty if the baryons differed from the electron in this simple way. However, experiments on the decay of protonium to $K_1^0 + K_1^0$ are strong evidence against this assumption. It seems that nature has reserved the positive sign in Eq. (12.1) for particles with integral spin and the negative sign for those with half-integral spin.

The square of the operator $U(IT)$ evidently represents the unit ray operator. Therefore it is a diagonal operator and commutes with all observables. According to Eq. (12.1), its restriction to an irreducible subspace is

$$
\begin{aligned}
[U(IT)]^2 &= \begin{pmatrix} 0 & \pm z \\ z^* & 0 \end{pmatrix} \begin{pmatrix} 0 & \pm z^* \\ z & 0 \end{pmatrix} \otimes [U_0(IT)]^2 \\
&= \pm \begin{pmatrix} z^2 & 0 \\ 0 & z^{-2} \end{pmatrix} \otimes [U_0(IT)]^2 \quad (|z|=1) .
\end{aligned}
\tag{12.2}
$$

If the above-mentioned connection between spin and the sign of $[U(I)]^2$ is adopted, the sign just cancels $[U_0(IT)]^2$ which is $+1$ for integral and -1 for half-integral spins. Then we have

$$
[U(IT)]^2 = \begin{pmatrix} z^2 & 0 \\ 0 & z^{-2} \end{pmatrix} \otimes 1 .
\tag{12.3}
$$

On many-body states $(n, \bar{n})$ of a given corpuscle, the operator $[U(IT)]^2$ has the value $z^{2(n-\bar{n})}$. It is convenient to define an operator Q with eigenvalues $n - \bar{n}$ on subspaces $(n, \bar{n})$, and to write

$$
[U(IT)]^2 = e^{iQ\alpha} ,
\tag{12.4}
$$

where α is arbitrary. Equation (12.4) summarizes our previous results: States with equal "generalized charge" $n - \bar{n}$ are mutually coherent, those with different generalized charge are mutually incoherent. Equation (12.4), which expresses incoherence as a consequence of gauge invariance, should be considered purely as a mnemonic device. In contrast to other authors (e. g., Yang and Tiomno [17]), we have not derived incoherences from assumed phases in the irreducible representers; we have rather assumed maximal coherence and adopted phases accordingly. The remaining non-trivial phases have been kept merely as reminders of residual incoherence which is due to physical prime principles, not to *ad hoc* assumptions.

13. Composition of different corpuscles

We have discussed many-body states of one kind of corpuscle. The bewildering variety of different particles must be reduced to some reasonably simple scheme if it is to be treated in a similar way.

We may disregard all unstable particles, since the present discussion concerns only rigorous symmetries and corpuscles should correspond to rigorously irreducible group realizations.

There remains a tremendous number of stable "particles", including chairs and tables. As first pointed out by HAAG [1], a systematic theory knows no distinction between "elementary" (i. e., small or light) and other stable particles.

Simplification of this confusing picture is achieved by considering − intuitively speaking − the largest possible number of particles as composites of a few simple particles. Since we are not concerned with dynamics, we must formulate this intuitive idea in terms of coherence.

A set of simple particles will be considered complete if every particle is coherent with some many-body state composed of simple particles. The bridge between this statement and the intuitive meaning of "composite particle" is the implicit belief that whenever the formation of a particle by collision of simple particles is not absolutely forbidden (by incoherence) it will indeed happen. A set of particles that appears to satisfy these requirements is: electron-positron, proton-antiproton, photon, and neutrino *. We will try to compose ray realizations of families of many-body states so as to make the largest possible number of them coherent.

At first sight, it would seen possible to compose the two irreducible subspaces (e, ē), (p p̄) so that, in accordance with Eq. (11.5), (e coh p̄) and (ē coh p), or vice versa. It is somewhat surprising that this possibility is eliminated on purely deductive grounds, i. e., that our general assumptions, together with the empirical list of primitive particles, are sufficient to derive or explain the existence of *two* generalized charges.

Assume first that maximal coherence between states of photons, electrons, and their antiparticles is established in the most obvious way. By choosing z in Eq. (12.1) equal for the two irreducible subspaces, we can obtain coherence between states with constant difference

$$n_e + n_{\bar{p}} - n_p - n_{\bar{e}} \, .$$

Each of the two subspaces (e, p) and (ē, p̄) would then be coherent with the vacuum. If there exist distinct stable particles (H, H̄) coherent with (e, p) and with (ē, p̄), respectively, then they must have two properties: (1) they must be equivalent representation spaces of $G'_{+\, \uparrow}$, and (2) they must be connected by space inversion. By T.7 we must then infer that the two particles H and H̄ are mutually incoherent (unless they have zero mass). However, by T.9, this is contrary to our assumption that they are both coherent with the vacuum. We can conclude, without

---

* Strictly speaking, the photon could be omitted, since it is coherent with the vacuum and with the proton-antiproton states.

further empirical evidence, that our tentative assumption of coherence between e and $\bar{\text{p}}$ and between $\bar{\text{e}}$ and p is not tenable, unless we made restrictive assumptions on the composite particles H, $\bar{\text{H}}$.

Accepting this additional incoherence, we look for those phase assignments which will maximize coherence. We may obtain coherence between many-body states consisting of p, $\bar{\text{p}}$, e, $\bar{\text{e}}$ if there is mutual coherence between subspaces with equal values of Q, Q', where

$$
\begin{align}
\text{e} - \bar{\text{e}} &= Q , \\
\text{p} - \bar{\text{p}} &= Q' .
\end{align}
\tag{13.1}
$$

As a convenient mnemonic device, operators Q, Q' with integral eigenvalues may be introduced, and the restrictions of the operator $[U(IT)]^2$ on the families of many-body states (e, $\bar{\text{e}}$) and (p, $\bar{\text{p}}$) may be set equal to

$$
\begin{align}
[U(IT)]^2 &= e^{iQ\alpha} \qquad [\text{on states } (n_e, n_{\bar{e}})] , \\
[U(IT)]^2 &= e^{iQ'\beta} \qquad [\text{on states } (n_p, n_{\bar{p}})] ,
\end{align}
\tag{13.2}
$$

where the eigenvalues are

$$
\begin{align}
Q &= n_e - n_{\bar{e}} , \\
Q' &= n_p - n_{\bar{p}} .
\end{align}
\tag{13.3}
$$

The residual coherence between the other many-body states are then expressed by the statement that the operator

$$
[U(IT)]^2 = e^{i(Q\alpha + Q'\beta)}
\tag{13.4}
$$

for any real α, β commutes with all observables. We remind ourselves again that this "gauge invariance" is purely a mnemonic device, all the statements implied by it having been proved independently.

The usual generalized charges are, of course, not our Q and Q', but Q and the baryon number $B = Q' - Q$. Clearly, any linear combination of the two generalized charges may be considered as a generalized charge, without changing the observable implications.

In our scheme, charge and baryon number are on the same footing. Any difference between them is dynamical, and not one of symmetry classification.

It is remarkable that there is no place in our scheme for other generalized charges, such as lepton numbers. It is not clear at present whether or not the selection rules which appear in weak interactions are superselection rules. In particular, it is not clear whether or not neutrinos with opposite helicity are mutually incoherent. According to our scheme, there is no reason for such an incoherence. Hence, we may expect the existence of linearly polarized neutrinos, as there are linearly polarized photons. This prediction might be considered as a possible test of the present theory, but I would be embarrassed if asked for the design of a Nicol prism for neutrinos.

Appendix 1

The first part of the statement: if one ray of $\mathscr{H}_i$ is mapped into $\mathscr{H}_j$, then all rays of $\mathscr{H}_i$ are, is proved as follows. If $f_1, f_2 \in \mathscr{H}_i$, there exists a ray f_3 "between" them in $\mathscr{H}_i$, i. e.,

$$d(f_1, f_3) < 2$$
$$d(f_2, f_3) < 2 .$$

If
$$Uf_1 \in \mathscr{H}_j, \ Uf_2 \in \mathscr{H}_k \quad (j \neq k) ,$$

then at least one of the distances, $d(Uf_1, Uf_3)$ or $d(Uf_2, Uf_3)$, must equal 2, since Uf_3 is in a subspace orthogonal to either $\mathscr{H}_j$ or $\mathscr{H}_k$. This contradicts Eq. (4.3).

The second part of the statement: the mapping is *onto* and not *into* $\bigcup \mathscr{H}_i$ (i. e., every ray in $\bigcup \mathscr{H}_i$ is an image point), follows from the fact that every ray operator $U(L)$ has an inverse since by Eq. (4.2)

$$U(L) \ U(L^{-1}) = 1 .$$

Appendix 2

Equations (4.7) and (4.8) can be simplified by using the structure of the isometries $\tilde{U}$ and hence of the semilinear operators U. If U is written as a matrix with elements U_{ik} which are semilinear operators on subspaces, so that U_{ik} is the restriction to $\mathscr{H}_i$ of U which maps $\mathscr{H}_i$ onto $\mathscr{H}_k$, then T.1 states that only one element U_{ik} in each is nonvanishing; and the unitary or antiunitary nature of U leads to the corresponding result for the rows; only one element U_{ik} in each row is different from zero. A matrix element may therefore be written as

$$U_{ik} = U_{ik} \ \delta\{k - f(i)\} . \tag{A2.1}$$

Consider a product $U D'$, where

$$D'_{ik} = d'_i \ \delta_{ik}; \ |d'_i| = 1 , \tag{A2.2}$$

then
$$U_{ik} \ \delta\{k - f(i)\} \ d'_k = d'_{f(i)} \ U_{ik} \ \delta\{k - f(i)\} . \tag{A2.3}$$

Let D'' have the matrix elements

$$D''_{ik} = d'_{f(i)} \ \delta_{ik} . \tag{A2.4}$$

We have then
$$U D' = D'' U . \tag{A2.5}$$

The right-multiplying operators D' in Eqs. (4.7) and (4.8) can always be replaced by left multipliers. We may then simplify these equations by writing

$$U(L) = \{D(L) \ \tilde{U}(L)\} \tag{A2.6}$$

and
$$U(L_1) \ U(L_2) = D(L_1, L_2) \ U(L_1 L_2) . \tag{A2.7}$$

Appendix 3

From

$$[U(I)]^2 = D = \begin{pmatrix} d_1 & 0 \\ 0 & d_2 \end{pmatrix}$$

$$(|d_1| = |d_2| = 1) ,$$

(A3.1)

we have

$$\alpha\,\beta = d_1 ,$$

$$\beta\,\alpha = d_2 ,$$

(A3.2)

or

$$d_1 = d_2 = d ; \quad \beta = d\alpha^+$$

$$[U(I)]^2 = d \begin{pmatrix} 1 & 0 \\ 0 & 1 \end{pmatrix} ; \quad (|d| = 1) .$$

(A3.3)

The general form of $U(L)$ $(L \in G_{+\uparrow})$ is

$$U(L) = \begin{pmatrix} u_1(L) & 0 \\ 0 & u_2(L) \end{pmatrix} .$$

(A3.4)

Consider an element $L \in G_{+\uparrow}$ and its transform $ILI^{-1} = ILI$ which is also a member of $G_{+\uparrow}$. The multiplication table of vector operators gives

$$U(I)\,U(L)\,U^{-1}(I) = D'(L)\,U(ILI) .$$

(A3.5)

Multiplication of two equations of this type gives

$$U(I)\,U(L_1)\,U(L_2)\,U^{-1}(I) = D'(L_1)\,D'(L_2)\,U(IL_1 I)\,U(IL_2 I) \quad \text{(A3.6)}$$

and by Eq. (5.6)

$$\pm U(I)\,U(L_1 L_2)\,U^{-1}(I) = \pm D'(L_1)\,D'(L_2)\,U(IL_1 L_2 I) . \quad \text{(A3.7)}$$

By comparing with Eq. (A3.6), one sees that

$$D'(L_1\,L_2) = D'(L_1)\,D'(L_2) ,$$

(A3.8)

that is, the two-dimensional unitary matrices D' form a unitary representation of the connected Poincaré group. However, it is known from representation theory that the only finite-dimensional unitary representation is

$$D'(L) = 1 .$$

(A3.9)

Equation (A3.5) with Eqs. (6.2), (A3.2), (A3.3), (A3.4), and (A3.9) reads

$$\begin{pmatrix} \alpha u_2(L)\,\alpha^+ & 0 \\ 0 & \alpha^+\,u_1(L)\,\alpha \end{pmatrix} = \begin{pmatrix} u_1(ILI) & 0 \\ 0 & u_2(ILI) \end{pmatrix} .$$

(6.4)

Appendix 4

Returning to Eq. (6.4), we attempt to find an operator α (on $\mathfrak{h}_1$) which satisfies it for the case $m \neq 0$ where, as noted before, we may assume $u_1 = u_2 = u$. We have then

$$\alpha^+\,u(L)\,\alpha = \alpha\,u(L)\,\alpha^+ = u(ILI) .$$

(A4.1)

Since $\alpha^+ = \alpha^{-1}$,

$$[\alpha^2, u(L)] = 0 \quad (L \in G_{+\uparrow}) \,. \tag{A4.2}$$

By Schur's lemma, α^2 must be a constant $c(|c| = 1)$.

In an irreducible representation, the operators $u(L)$ generate the full v. Neumann algebra of bounded operators. It is easy to see that the operator α which induces the automorphism

$$u(L) \rightarrow u(ILI)$$

is unique up to a factor of unit modulus. Therefore, α must be equal to the representative $U_0(I)$ which is determined in the conventional theory of coherent representations, since it induces the same automorphism. We have

$$U(I) = \begin{pmatrix} 0 & z \\ z & 0 \end{pmatrix} \otimes U_0(I) \quad (|z| = 1) \,. \tag{A4.3}$$

The class of realizations of this type has a one-to-one correspondence to the irreducible representations of $G_{+\uparrow}$ or $G'_{+\uparrow}$ with $m \neq 0$. The realizations are irreducible, for if there were a proper invariant subspace $\mathfrak{h}' \leq \mathfrak{h}_1$, this would contradict the irreducibility of the vector rerpesentation $u(L)$ $(L \in G_{+\uparrow})$, and if the invariant subset were one of the coherent subspaces $\mathfrak{h}_i$, then $U(I)$ would carry it into the other coherent subspace. Conversely, if $u(L)$ $(L \in G_{+\uparrow})$ are reducible representations of $G_{+\uparrow}$, clearly proper invariant subspaces can be found. Thus, we have found the complete set of irreducible realizations with incoherence.

We now wish to add the time-reversal element T to the orthochronous group. Since the operator $U(T)$ leaves the coherent subspaces invariant, the problem of finding a representative $U(T)$ is the same as is the theory of coherent representations, and its solution is well known. Finally, $U(IT)$ can be obtained by multiplying $U(I)\,U(T)$ by a diagonal matrix. The general set of vector operators is:

$$U(L) = D(L) \begin{pmatrix} 1 & 0 \\ 0 & 1 \end{pmatrix} \otimes U_0(L) \quad (L \in G_{+\uparrow})$$

$$U(I) = D(I) \begin{pmatrix} 0 & 1 \\ 1 & 0 \end{pmatrix} \otimes U_0(I)$$

$$U(T) = D(T) \begin{pmatrix} 1 & 0 \\ 0 & 1 \end{pmatrix} \otimes U_0(T)$$

$$U(IT) = D(IT) \begin{pmatrix} 0 & 1 \\ 1 & 0 \end{pmatrix} \otimes U_0(IT) \,.$$

References

[1] HAAG, R.: Kgl. Danske Videnskab. Selskab, Mat-fys. Medd. 29, No. 12 (1955)
[2] MELVIN, M. A.: Revs. Modern Phys. 32, 477 (1960)
[3] LANDAU, L.: Nuclear Phys. 3, 127 (1957)
[4] SALAM, A.: Nuovo cimento 5, 299 (1957)
[5] EKSTEIN, H.: Phys. Rev. 120, 1917 (1960)
[6] — Nuovo cimento 23, 606 (1962)

[7] Wightman, A. S.: Nuovo cimento Suppl. **14**, 81 (1959)
[8] Segal, I. E.: Mathematical Problems of Relativistic Physics (Ann. Math. Soc. Providence, 1963)
[9] Mackey, G. W.: Mathematical Foundations of Quantum Mechanics. New York: W. A. Benjamin 1963
[10] Wick, G. C., A. S. Wightman, and E. P. Wigner: Phys. Rev. **88**, 101 (1952)
[11] von Neumann, John: Mathematical Foundations of Quantum Mechanics, p. 313. Princeton: Princeton University Press 1955
[12] Haag, R., and D. Kastler: J. Math. Physics **5**, 548 (1964)
[13] Reference [9], p. 82
[14] Bargmann, V.: Ann. Math. **59**, 1 (1954)
[15] Shirokov, Iu. M.: Soviet Phys. JETP **6** (33), 604 (1957)
[16] Cook, J. M.: Trans. Am. Math. Soc. **74**, 233 (1953)
[17] Yang, C. N., and J. Tiomno: Phys. Rev. **79**, 495 (1950)

Prof. Hans Ekstein

Argonne National Laboratory D 203
Argonne/Illinois (USA)